Systematic Approach
to Evaluation of
# MOUSE
# MUTATIONS

# Systematic Approach to Evaluation of
# MOUSE MUTATIONS

edited by

## John P. Sundberg, D.V.M., Ph.D.
## Dawnalyn Boggess, B.S.

The Jackson Laboratory
Bar Harbor, Maine

**CRC Press**
Taylor & Francis Group
Boca Raton London New York

CRC Press is an imprint of the
Taylor & Francis Group, an **informa** business

CRC Press
Taylor & Francis Group
6000 Broken Sound Parkway NW, Suite 300
Boca Raton, FL 33487-2742

First issued in paperback 2019

ISBN-13: 978-0-8493-1905-1 (hbk)
ISBN-13: 978-0-367-39958-0 (pbk)

**Library of Congress Cataloging-in-Publication Data**

Systematic approach to evaluation of mouse mutations / editors, John P. Sundberg and Dawnalyn Boggess
    p.  cm.
    Includes bibliographical references and index.
    ISBN 0-8493-1905-6 (alk. paper)
    1. Mice – Genetics. 2. Genetics – Animal models. 3. Mice as laboratory animals. I. Sundberg, John P. II. Boggess, Dawnalyn.
QH432.S97 1999
576.5′.49—dc21                                           99-14308
                                                             CIP

Library of Congress Card Number 99-14308

**Visit the Taylor & Francis Web site at
http://www.taylorandfrancis.com**

**and the CRC Press Web site at
http://www.crcpress.com**

# About the Editors

**John P. Sundberg, D.V.M., Ph.D.,** is head of the pathology program and a senior staff scientist at The Jackson Laboratory in Bar Harbor, Maine. Dr. Sundberg graduated in 1973 from the University of Vermont with a B.S. degree in Animal Science (*summa cum laude*) and obtained his D.V.M. degree in 1977 from Purdue University School of Veterinary Medicine, West Lafayette, Indiana. Following a brief period in private practice, Dr. Sundberg earned a Ph.D. degree in comparative pathology and virology in 1981 from The University of Connecticut in Storrs. Dr. Sundberg served as an assistant professor at the University of Illinois College of Veterinary Medicine from 1981 to 1986. In 1986 he assumed his present position.

Dr. Sundberg is a diplomate of the American College of Veterinary Pathologists. He is a member of the American Veterinary Medical Association, American Association of Laboratory Animal Science, Society for Investigative Dermatology, and the Hair Research Society.

Dr. Sundberg has received research grants from the National Institutes of Health, American Cancer Society, National Alopecia Areata Foundation, and private industry. He has published over 180 research and clinical papers, 50 technical bulletins, 95 book chapters, and two books. His current major research interests relate to mouse mutations as models of human and domestic animal dermatological diseases, the comparative pathology and molecular evolution of papillomaviruses, and spontaneous diseases of inbred laboratory mice.

**Dawnalyn Boggess, B.S.,** is a professional research assistant at The Jackson Laboratory in Bar Harbor, Maine. Ms. Boggess graduated in 1989 from the University of Maine with a B.S. degree in Animal Science. She has worked as a research assistant and laboratory manager in Dr. Sundberg's Dermatology Research Program since February 1990.

Ms. Boggess is a licensed veterinary technician and is certified by the American Association for Laboratory Animal Science (AALAS) as a Laboratory Animal Technologist. She is author or co-author of more than 20 research papers and five book chapters, three of which are included in this book.

# Contributors

**Charles C. Bascom, Ph.D.**
The Procter & Gamble Company
Miami Valley Laboratories
Cincinnati, Ohio

**Lesley S. Bechtold, M.S.**
The Jackson Laboratory
Bar Harbor, Maine

**Dawnalyn Boggess, B.S.**
The Jackson Laboratory
Bar Harbor, Maine

**Terrie L. Cunliffe-Beamer,
D.V.M., M.S.**
The Jackson Laboratory
Bar Harbor, Maine

**Muriel T. Davisson, Ph.D.**
The Jackson Laboratory
Bar Harbor, Maine

**Diane S. Keeney, Ph.D.**
Department of Cell Biology
Vanderbilt University
Nashville, Tennessee

**Lloyd E. King, Jr., M.D., Ph.D.**
Dermatology Clinic
The Vanderbilt Clinic
Nashville, Tennessee

**David N. Larkins**
The Jackson Laboratory
Bar Harbor, Maine

**Brian J. Limberg, B.S.**
The Procter & Gamble Company
Miami Valley Laboratories
Cincinnati, Ohio

**Gregory Martin, M.S.**
The Jackson Laboratory
Bar Harbor, Maine

**James Miller**
The Jackson Laboratory
Bar Harbor, Maine

**Xavier Montagutelli, D.V.M., Ph.D.**
Institut Pasteur
Unité de Genétique des Mammiferes
Paris, France

**Melissa J. Relyea, B.S.**
The Jackson Laboratory
Bar Harbor, Maine

**John J. Sharp, Ph.D.**
The Jackson Laboratory
Bar Harbor, Maine

**Richard S. Smith, M.D., D. Med. Sci.**
The Jackson Laboratory
Bar Harbor, Maine

**Beth A. Sundberg, M.S.**
The Jackson Laboratory
Bar Harbor, Maine

**John P. Sundberg, D.V.M., Ph.D.**
The Jackson Laboratory
Bar Harbor, Maine

**Jerrold M. Ward, D.V.M., Ph.D.**
NCI Veterinary Pathology
Frederick, Maryland

# Contents

# Introduction

The 1990s have been called the decade of the mouse, but it appears that mammalian genetics research based on inbred laboratory mice will be one of the mainstays of biomedical research into the 21st century. Identification of spontaneous mouse mutations provides tools to discover novel genes. Partial or total inactivation of known genes, in the form of targeted mutagenesis (so-called "knock out" mutations), or overexpressing genes in specific anatomic structures (transgenesis) provides another set of tools to help us understand gene function. The real value of all of these approaches depends upon how carefully we are able to study the biology of the mice produced. The rush to publish continues to fill the literature with incomplete evaluation of mutant mice, making many of the interpretations of questionable value. Interaction between molecular biologists, who in reality, have become specialized chemists; traditional biologists and veterinarians, who understand the medical and pathological aspects of the mouse; and physicians, who understand the human side of the equation, all working together, yield the most thorough evaluation and interpretation of these types of investigations. These types of collaborations cannot always be forged. Even when they can, the group is not always experienced enough to do a thorough and accurate job. The purpose of this book is to provide a systematic approach for evaluating mouse mutations. This approach evolved through years of developing collaborations with other scientists to fill the holes in one's individual intellectual shortcomings.

The mouse is just a small version of any of a number of mammals that veterinarians and physicians work with daily. Therefore, most of the methods described here are, in a sense, scaled-down versions of standard operating procedures for diagnostic medicine and biomedical research. This book focuses on the laboratory mouse to give the reader specific methods for working with these animals. Resources, both for reagents and the mice themselves, are very specialized. Nomenclature for pathological descriptions is standard for all medical systems. However, mouse genetic nomenclature is very specialized. Therefore, integrating all these types of basic issues should provide the reader with a single resource to begin sophisticated biomedical studies.

# 1 Colony Establishment

*Dawnalyn Boggess, Terrie L. Cunliffe-Beamer, and John P. Sundberg*

## CONTENTS

## I. INTRODUCTION

Inbred laboratory mice have long been the species of choice for biomedical research. This is due to (1) the relatively low cost of maintaining a colony, (2) the genetic homogeneity of inbred mice, (3) the availability of numerous reagents for very specialized testing, (4) the high degree of genetic homology between mice and humans, and (5) the many sophisticated genetic tools available for working with mice. Mutations in laboratory mice may occur naturally, in part due to inbreeding, or may be induced using transgenic and gene targeting (called "knock-out") technology, radiation, mutagenic chemicals, or viruses.

## II. DIFFERENTIATING A MUTATION FROM AN INFECTIOUS DISEASE

Naturally occurring mutations (phenotypic deviants) are usually first observed by the animal care technicians who maintain the colonies. Once identified, these phenotypic deviants are brought to the attention of the principal investigator or veterinary staff. At this point, it is necessary to determine whether the observed abnormality is due to (1) an infectious disease that could become a serious problem for the mouse colony and perhaps the entire facility, (2) a noninfectious disease due to an environmental problem, (3) a common background disease characteristic of that particular inbred strain, or (4) a potential new mutation (Figure 1.1). Descriptions of common murine infectious diseases and strain-specific diseases are published in

## Identification of a Phenotypic Deviant

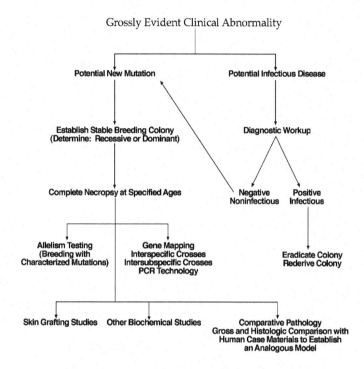

**FIGURE 1.1** Flow pattern to differentiate a potential new mutation from an infectious disease outbreak. (From Sundberg, John P. et al., Systematic approach to evaluation of mouse mutations with cutaneous appendage defects, in Chuong, C-M. Ed., *Molecular Basis of Epithelial Appendage Morphogenesis*, R.G. Landes Co., Austin, TX, 1998, 422. With permission.)

detail elsewhere.[1-4] Once an infectious disease or environmentally induced condition has been ruled out, or if only one mouse or a small group of mice from one litter are observed to be abnormal, it would be reasonable to consider these mice to be phenotypic deviants or potential mutant mice.

To aid in determining whether a phenotypic deviant is due to a genetic mutation or the result of a random event, such as a spontaneous disease or accidental developmental defect, one should look for the following: (1) recurrence of the phenotype in mice from subsequent litters of the same breeder pair, (2) occurrence of the same phenotype in mice from different breeder pairs within the same pedigree or family line, (3) occurrence of more than one mouse with the abnormal phenotype in the same litter, or (4) presence of both normal and affected mice in the same litter.

Mice derived from transgenic or targeted mutagenesis technology are generally designed to enable a researcher to study the function of a specific gene. However, as was the case with the *Fgf5* null, an allele of the spontaneous angora mutation,[5] the anticipated result is not always what is achieved. Knockout of the *Fgf5* gene merely caused growing of long hair rather than neonatal lethality. For this reason, it is as important to evaluate these mutant mice in the same systematic way you would a naturally occurring mutation.

Some of the most easily observed phenotypic differences are in coat color, coat texture, and certain abnormalities of the skin. These can be due to hair or pilosebaceous defects but may also affect nail growth, another cutaneous appendage. Other types of phenotypic deviants also observable by clinical examination include: defects of the limbs and/or tail, which may be due to bone or muscle abnormalities; behavioral abnormalities, such as unsteadiness of gait or posture, which could be caused by a neurological or muscle defect; or changes in color of mucous membrane, which can indicate anemia, or cardiac or liver defects, depending on the color. Moribund mice may have abnormalities in any organ system.

## III. IDENTIFYING PHENOTYPE

It is important, when establishing a colony of potentially mutant mice, to examine all mice as soon after birth as possible and at regular intervals (at least every other day during the first two weeks) thereafter. Defects may be observable at birth that make the mutants easy to identify (as is seen in flaky skin [fsn] mice, which have a mild anemia at birth,[6] or juvenile alopecia [jal] mice, which have easily identifiable abnormalities of the vibrissae as early as 2 to 3 days of age).[7] Other defects may develop as the mouse ages, as in harlequin ichthyosis (ichq) mice, which are normal until 5 days of age when they develop thick, scaling skin, then die by 10 to 12 days of age,[8] or matted (ma) mutants, in which the hairs form matted clumps, such that mutants can be identified at 2 to 4 weeks of age,[9] and chronic proliferative dermatitis (cpdm), in which the skin lesions become apparent at 4 to 6 weeks of age.[10] Some defects may even disappear as the animal ages, as with flaky tail (ft), in which the skin defect is recognizable by 2 to 4 days of age and has usually disappeared and mutant mice appear normal by 14 to 15 days of age.[11]

## IV. COLONY ESTABLISHMENT

Establishing a colony of mice carrying the suspected mutation is best done, of course, by breeding the deviant mouse. If the deviant dies or is sterile, obtain the parents of the deviant pup or litter and allow them to continue producing litters. If the parents are unavailable or are too old to produce more litters, the siblings of the affected mouse can be mated together to produce more litters. If any of the offspring of the parental or sibling matings produce more of the phenotype in question, then it is possibly caused by a genetic mutation.

When you have determined that the phenotype is reproducible (transmitted), you can then expand the colony. At this time you will also want to do a systematic diagnostic necropsy of at least one mutant and one normal littermate. This is to further rule out the possibility of infectious disease, as well as to attempt to identify the disease processes that are causing the phenotype of the deviants. This will allow you to better direct the characterization and maintenance of the colony.

The colony can be expanded by simple brother × sister matings. If the mutation is recessive and mutants cannot be bred, matings of this type would be considered unknown test matings. If any of these matings produce at least one affected offspring in a litter, the animals in that mating would be considered tested breeders. Matings are usually maintained until at least 18 to 20 pups are born, with no deviants found, before being deemed a negative test mating. Nonproductive breeders and breeders that produce no affected offspring should be remated to tested breeders or removed from the colony. If the number of mice available is small, exchange males among females. If recessive mutant mice can breed, it is a good idea to breed homozygous (*m/m*) mutants to heterozygous (+/*m*) or wildtype (+/+) litter mates. With these types of matings, you are assured of getting litters that will either have some homozygous, affected mice and some heterozygous, normal mice or, if none are affected, all heterozygous mice. Recessive and dominant mutant mice can also be maintained by homozygous matings (*m/m* × *m/m* or *M/M* × *M/M*, respectively), however, no unaffected mice will be born. There are some mutations, which, due to the phenotype, make the females poor mothers. Lethal milk mutation is an example of this.[12] For this reason, it is usually best to mate mutant males to normal (heterozygous or wildtype) females. When using the +/*m* × +/*m* breeding scheme for vigor, mate *m/m* × +/+ in alternate generations to produce known +/*m* mice. Dominant mutations can be maintained by mating heterozygotes (*M/+*) to +/+ siblings or +/+ mice of the same inbred genetic background as that on which the mutation occurred.

If some or all newborn pups in litters consistently die or fail to thrive, fostering pups is a way to determine if the problem is inherent in the pups or is related to maternal care or lactation. Fostering is generally done at birth or within 2 to 3 days of birth to have the best chance of success. The foster mother should have a history of successfully rearing virtually all pups in one or more litters and come from a strain that has good maternal care,[13] but should not be at the end of her reproductive lifespan. Lactation may start to fail in female mice at the end of their reproductive lifespan, and this can confound results of fostering studies. There should be a coat color difference between fostered and natural pups to facilitate observation of the fostered pups. If the litter to be fostered is especially large, two foster mothers should be considered. The foster mother's litter and the litter to be fostered should be age matched and size matched as closely as possible for the best chance of success. Foster females sometimes reject pups that are different in size from their own. Also, composition of mouse milk changes during lactation[14,15] and no doubt influences the growth of the pups. Lactation peaks at 12 to 13 days postpartum and then rapidly declines.[16] Milk production is almost ceased, and mammary glands are involuting by the time a litter is around 21 days of age.

Successful hand-rearing of newborn mice has been described.[17,18] Newborn mice cannot regulate body temperature and must be maintained in a warm environment. They must be hand-fed small amounts every two hours until they are two weeks of age. At that time, soft diet and milk can be placed in open containers. Newborn mice do not urinate or defecate unless the perineal region is stimulated. The mother provides this stimulation when she licks her pups. After feeding, the perineal regions of hand-reared pups need to be cleansed with warm water and a soft swab to induce urination and defecation.

Another method of expanding and maintaining the colony if the mutant mice have reduced viability, or if the mutation was originally found on a segregating (outbred) background, is to create a congenic strain by crossing mice carrying the mutation (donor strain) with mice of another inbred strain (recipient strain). By selecting mice homozygous (if fertile) or heterozygous for the mutation and crossing these animals to the recipient strain and then selecting affected offspring to backcross to the recipient strain, it is possible to "transfer" the mutation onto the selected inbred strain, thereby creating a new congenic strain.[19,20] For a recessive mutation, mice must be intercrossed at alternate generations to recover homozygotes for the next backcross or to identify +/m carriers. Each backcross to the recipient strain is termed one backcross (N) generation, with the first cross to the recipient strain counting as N1. When alternative generation backcross–intercrosses are required, only the backcross generations are counted (termed NE, backcross equivalent generations) to determine percent recipient genome. A strain is considered congenic at N10 or NE generations, because it is statistically >99% recipient genome. One must be aware that the chromosomal segment around the mutation being transferred is likely to be of donor origin. Creation of the congenic strain and the subsequent inbreeding done after the congenic is created can make profound changes to the mutant phenotype compared to that observed in the original strain or stock. Therefore, it will be important to have collected tissues for histology from animals with the original mutation for comparison with tissues from the congenic mice to verify that the original phenotype was not lost or significantly changed. It is critical to keep accurate pedigree/breeding records whatever method is chosen to maintain the new mutant strain to accurately track the breeding history and genetic content of the strain.

## V. OVARIAN TRANSPLANTATION

Ovarian transplantation is another strategy to successfully breed certain kinds of mutant, transgenic, or knock-out mice. Ovarian transplantation should be considered when (1) the mutant ovary contains relatively normal numbers of functional oocytes that are capable of normal maturation and (2) gonadotrophin deficiency, physical abnormality, or early mortality prevent the mutant mouse from conceiving, mating, or sustaining a pregnancy. Ovary transplantation will not repair absence of germ cells or maturation defects that are inherent in the germ cell.

Ovary transplant requires a donor mouse and one or more recipient mice. Ovaries from postweanling mice are usually divided in half, and one-half of an ovary is unilaterally transplanted into each of four recipients. Ovaries from very young, preweanling mice are usually transplanted whole. Donor and recipient mice must be compatible at major and minor histocompatibility alleles. An F1 hybrid between two inbred strains can serve as ovary transplant recipient for an ovary donor from either parental inbred. An F1 hybrid can also serve as the ovary transplant recipient for an ovary donor from the F2 generation.[20] Alternatively, a homozygous severe combined immunodeficient (*Prkdc^scid^/Prkdc^scid^*) mouse or a mouse with similar severe immunodeficiency can be used as a recipient. Mice with severe immunodeficiency require housing systems and husbandry procedures that will protect them from exposure to primary pathogens of mice as well as opportunistic pathogens of mice. In nonbarrier facilities, oral administration of sulfamethoxazole–trimethoprim suspension (200 mg sulfamethoxazole, 40 trimethoprim per 5 ml) may be considered to reduce morbidity and mortality due to *Pneumocystis carinii* infection in severely immunodeficient mice, such as *Prkdc^scid^/Prkdc^scid^* mice. One regimen is to make a 2% solution from a commercially available suspension that contains 200 mg sulfamethoxazole, 40 mg trimethoprim per 5 ml and water. The 2% solution is provided in lieu of drinking water for three days during each week.[21-23] Decision to administer sulfamethoxazole–trimethoprim should be made in consultation with the institution's clinical veterinarian because long-term administration of antibiotics is not without potential side effects. *Prkdc^scid^/Prkdc^scid^* mice undergoing ovary transplant using aseptic and atraumatic techniques do not require antibiotics to prevent postoperative infections.[24]

Recipient mice should be between 4 and 7 weeks of age. Response of mice less than 4 weeks of age to injectable anesthetics is not predictable. Compared to younger mice, recipient mice over 7 weeks of age are more likely to have increased abdominal fat and increased problems with hemorrhage, depending on the stage of the estrus cycle. Older mice have been successfully used as recipients, but "the surgeon" must be prepared for increased fat and hemorrhage. The recipient mouse should carry a dominant coat color gene or biochemical marker that permits identification of her biological pups in the event that excision of the recipient's ovaries is incomplete. White-bellied agouti (gene symbol $A^W$) is often used for this purpose, because this allele is dominant to virtually all other alleles at the agouti locus. If homozygous or heterozygous mutant mice can be easily identified by DNA test, e.g., polymerase chain reaction, Southern blot, biochemical marker, or obvious phenotype, a coat-color marker is less critical. However, even individuals who are experienced and very skilled at ovary transplantation will occasionally leave a small piece of recipient ovary. Pups from both transplanted ovary and the small remnant of the recipient's ovary can be born in the same litter. On occasion, the first few litters will contain only pups from transplanted ovary, and pups from a remnant of recipient ovary may not appear until later litters are born.

Donor mice should be 3 to 8 weeks of age, if at all possible. Ovaries can be transplanted from preweanling mutant mice if the mutation causes mortality prior to weaning, but these young ovaries are small. Ovaries can also be transplanted from aged mice, but the probability of successful transplant is reduced, because older mice have fewer numbers of oocytes than younger mice.

Ovary transplantation is a major surgical procedure and was initially described in the 1940s. Since then, several descriptions of the technique have been published.[25-31] A dissecting microscope with a working distance between 5 and 10 inches, depth of field of approximately 0.75 inch on low-power magnification, and field diameter around 1 inch on low-power magnification is required. The microscope's range of magnification should be between 4× and 25×. Aseptic and atraumatic surgical practices must be observed to maximize success. Mastery of ovarian transplantation requires practice. Skilled personnel can expect to achieve 80% or higher successful transplants.

The transplant is usually made on the recipient's left side, and ovariectomy only is performed on the recipient's right side. This is done because surgical exposure of the left ovary is easier due to its more caudal location in the abdomen. However, it is possible to transplant to the right side only, or to both sides if desired. One half of a donor ovary is transplanted into the ovarian bursa of the recipient following ovariectomy. If donor ovaries are very small, the entire ovary can be transplanted. Thus, ovaries from one donor can be transplanted to one, two, or four recipients, depending on whether a whole ovary is transplanted bilaterally into a recipient, or a whole ovary is transplanted unilaterally into two recipients, or one half an ovary is transplanted bilaterally into two recipients, or one half an ovary is transplanted unilaterally into four recipients.

Donor and recipient surgery packs are required. The donor pack consists of 3.5" or 4.5" sharp/sharp microdissecting scissors and thumb forceps for skin, 3.5" or 4.5" sharp/sharp microdissecting scissors and forceps for abdominal wall, two pairs of Dumont watchmaker forceps (straight or curved depending on personal preference), and 3.5" sharp/sharp microdissecting scissors or Vannas, DeWecker, or similar scissors. A small sterile Petri dish or watch glass and sterile buffered saline are also required.

The recipient pack consists of a drape, $2 \times 2$" Nu-Gauze sponges, very small surgical spears or toothpick swabs,[32] 3.5" or 4.5" sharp/sharp microdissecting scissors and microdissecting serrated angular forceps (0.8 mm tip) or microdissecting $1 \times 2$ teeth forceps for skin, 3.5" or 4.5" sharp/sharp microdissecting scissors and microdissecting $1 \times 2$ teeth forceps for abdominal wall, and two pairs of Dumont watchmaker forceps (straight or curved), microdissecting Vannas, DeWecker, or similar scissors, Baby Derf or similar needle holder, and a wound clip applicator and wound clips if desired for skin closure. 5–0 Dexon or similar absorbable suture with a CE-2 or similar preattached needle is used to close the incision in the abdominal wall.

The donor mouse is euthanized by carbon dioxide or cervical dislocation. The carcass is soaked with disinfectant (70% alcohol or quaternary ammonium) and placed on a paper towel. In the rare event that the donor mouse is not to be euthanized, procedures for survival surgery and ovariectomy are followed. Ovaries are excised from the euthanized donor mouse, using sterile instruments and aseptic technique. Either of two alternative approaches may be used to excise ovaries from the donor.

*Excision of donor ovaries. Technique 1:* The donor mouse is placed on its abdomen. The skin over the lumbar region is pulled apart in a medial–lateral direction with the fingers; one piece of skin is pulled cranially and the other piece of skin is pulled caudally to uncover the lumbar muscles and abdominal wall. Alternatively, sterile scissors are used to make an incision perpendicular to the vertebral column crest approximately midway between the iliac crest and the last rib. Then the skin incision is rotated laterally beyond the ventral edge of the lumbar muscles to expose the muscles of the abdominal wall.

A 3- to 6-mm incision is made in the abdominal wall. The incision in the abdominal wall is made between segmental blood vessels approximately 4 to 5 mm caudal to the last rib. Sterile forceps and scissors are used to incise the left abdominal wall approximately midway between last rib and an imaginary line perpendicular to the iliac crest. A second pair of sterile forceps is used to grasp and exteriorize the left ovarian fat pad. The ovary and part of the fat pad are torn free of mesenteric attachments and placed in a dish of cold sterile PBS or tissue culture media. The second ovary is located, excised in the same manner, and placed in the same dish of cold sterile PBS or media. The dish and ovaries are placed under a dissecting microscope. Each ovary is dissected free of ovarian bursa and fat pad, using Dumont forceps and fine pointed scissors. Normal size ovaries are cut in half.

*Excision of donor ovaries. Technique 2:* The donor mouse is placed on its back. The skin over the abdomen is pulled apart in a medial–lateral direction with the fingers. One piece of skin is pulled cranially and the other caudally to uncover the abdominal wall. Alternatively, sterile scissors are used to make an incision that starts at the pubic symphysis, turns laterally toward the left side of the mouse, and ends near the xiphoid cartilage. Then the skin is bluntly dissected away from the abdominal wall. The long flap of skin is folded toward the right side of the mouse.

A second set of sterile scissors and forceps is used to incise the abdominal wall from the pubic tendon toward the xiphoid cartilage. One uterine horn is elevated out of the incision with another pair of sterile forceps and the uterine horn is followed cranially until the ovary is exteriorized. The ovary and part of the fat pad are torn free of mesenteric attachments and placed in a dish of cold sterile PBS or tissue culture media. The second ovary is located, excised in the same manner, and placed in the same dish. The dish and ovaries are placed under a dissecting microscope. Each ovary is dissected free of ovarian bursa and fat pad, using sterile Dumont forceps and fine pointed scissors. Normal size ovaries are cut in half.

*Preparation of recipient:* The recipient mouse is anesthetized. Fur on the back is plucked (most adult mice have prolonged telogen-phase hair follicles; hairs are loose, easily removed, and removal is painless) or clipped from an area bounded cranial–caudal by the thoraco–lumbar junction and iliac crests, and dorsal–ventral by lateral edges of the lumbar muscles. Loose hair is picked up with an adhesive tape. Skin is disinfected with 70% alcohol or similar presurgical antiseptic using a sterile swab. Application of 70% alcohol starts in the center of the clipped area and works outward in ever-widening circles. The swab is discarded after it is used to spread fur away from the clipped area. Three applications of alcohol are made, using a new sterile swab each time. Alternatively, two applications of 70% alcohol and one application of surgical iodine may be used.

The recipient mouse is placed in ventral, tending toward right lateral, recumbent position. A right-handed surgeon should have the mouse's head pointed toward the surgeon's left hand and the mouse's body parallel to the surgeon's body. A left-handed surgeon should have the mouse's head pointed toward the surgeon's right hand. These adjustments in position of the mouse are necessary to make it easier for the surgeon to incise the ovarian capsule and assure that the oviduct is out of the way. A surgical drape is positioned over the mouse.

A 0.5- to 1-cm skin incision can be made perpendicular to the vertebral column crest approximately midway between the iliac crest and the last rib, using sterile scissors and forceps. These instruments are used only on skin. The skin incision is rotated laterally beyond the ventral edge of the lumbar muscles to expose the muscles of the left abdominal wall. Or a 5- to 6-mm skin incision can be made perpendicular to the long axis of the body starting midway between the iliac crest and the last rib at the ventral edge of the lumbar muscles and extending over the abdominal muscles.

A 3- to 6-mm incision is made in the left abdominal wall, using a second pair of scissors and forceps. These instruments are used only on the abdominal wall. The incision in the abdominal wall is made between segmental blood vessels approximately 4 to 5 mm caudal to the last rib. The incision is parallel to the rib extending from the ventral edge of the lumbar muscles toward the ventrum.

A third pair of sterile forceps is used to grasp and exteriorize the left ovarian fat pad. This pair of forceps is used only to manipulate the fat pad. The fat pad is positioned on a 2 × 2" Versalon sponge or similar sterile surface so that the ovary is facing the surgeon and the oviduct is toward the surgeon's left hand. If desired, the ovarian fat pad can be held in position, using a Schwartz temporary clip or baby Diffenbach serrefine. The ovarian capsule is incised at the junction of capsule and fat pad using Vannas, DeWecker, or Noyes ultra micro scissors. The incision in the ovarian capsule starts to the right of the oviduct at approximately 6 o'clock, extends cranially through 3 o'clock, and ends near 12 o'clock. Dumont forceps are used to slide the capsule off the ovary and place the capsule on top of the oviduct. One pair of Dumont forceps is placed under the ovary and used to clamp the ovarian blood vessels. A second pair of Dumont forceps is placed on top of the first pair and used to shear the ovary off the ovarian blood vessels. The first pair of Dumont forceps remains clamped on the ovarian blood vessels to prevent bleeding. The excised recipient ovary is discarded on a sterile surface, using the second pair of forceps, and that pair of forceps is used to pick up a piece of donor ovary. If bleeding is observed from the ovarian vessels, the Dumont forceps are removed, and hemorrhage is controlled using direct pressure and a surgical spear or a fine-tipped, cotton-wrapped toothpick. If bleeding is not observed, the piece of donor ovary is placed on top of the ovarian blood vessels as the first Dumont forceps are loosened and removed. The ovarian capsule is located near the oviduct, retracted over the donor ovary, and tucked back to its original position, using one or both pairs of Dumont forceps. If bleeding is observed from ovarian vessels after the donor ovary is in place, it may be controlled by direct pressure on the ovarian capsule. Alternatively, it may be necessary to retract the capsule and temporarily remove the donor ovary in order to apply direct pressure and control hemorrhage. Dumont forceps must be handled carefully to avoid tearing the ovarian capsule or accidentally damaging the oviduct.

If used, the temporary clip is removed from the fat pad. The fat pad is returned to the abdominal cavity, using the third pair of forceps. The "abdominal wall" forceps are used to approximate the edges of the incision in the abdominal wall. This incision is sutured with 5–0 or 6–0 absorbable suture with swaged-on needle.

If the skin incision was made in the left abdominal wall, it is closed with suture, wound clips, or tissue adhesive. If the skin incision was made in the dorsal lumbar area, it is rotated dorsally to its original position and then rotated laterally beyond the ventral edge of the lumbar muscles to expose the muscles of the right abdominal wall. The mouse is rotated slightly to a ventral, tending toward left lateral, position. A 3- to 6-mm incision is made in the right abdominal wall, using a second pair of "abdominal wall" scissors and forceps. The right ovarian fat pad is exteriorized. The usual procedure is to perform ovariectomy only and not transplant a piece of donor ovary to the right ovarian bursa because the right ovary has a shorter pedicle and is more difficult to position. The right ovary is excised using one Dumont forceps to crush ovarian blood vessels as they enter the base of the fat pad near the oviduct, and a second Dumont forceps is used to shear off the ovarian capsule, ovary, and piece of the oviduct. Alternatively, a transplant can be made to the right ovarian bursa, using the technique described above.

The fat pad is returned to the abdominal cavity, using the third pair of forceps. The "abdominal wall" forceps are used to approximate the edges of the incision in the abdominal wall. This incision is sutured with 5–0 or 6–0 absorbable suture with swaged-on needle. The skin incision is rotated dorsally to its original location and closed with sutures or wound clip. Tissue adhesive may be used in combination with sutures for skin closure.

There are several technical reasons for ovary transplant failure. If the incision in the capsule is made in the middle of the capsule or the capsule is accidentally torn, the remaining bursa may not be large enough to retain the transplanted ovary after it is returned to the abdominal cavity. In this case, the transplanted ovary may slip outside of the bursa and attach to the fat pad or be lost in the abdominal cavity. If the cut in the capsule is made across the center of the capsule, adhesions from the scarring in the capsule to the ovary may occur, and this can interfere with transit of oocytes to the infundibulum. If the incision is made in the fat pad instead of the junction of capsule, there is increased risk of hemorrhage and the capsule cannot be entered to expose the ovary. Ovaries transplanted into the fat pad are not functional. If hemorrhage from ovarian blood vessels is not controlled, blood can push the transplanted ovary out of the capsule into the abdominal cavity. If the recipient's ovaries are not completely excised, they can recover and produce functional oocytes. If the oviduct is accidentally pinched with the Dumont forceps, an unwanted "tubal ligation" may occur. If the incision in the abdominal wall is not sutured, the ovary and oviduct or other viscera can herniate through the incision, creating a subcutaneous hernia.

## VI. SETTING UP A COLONY FOR GENE MAPPING

Another important aspect of mutant mouse evaluation and colony management is the setting up of matings for genetic mapping purposes. If the mutant you are working with resembles a mutation that has already been characterized and mapped, you can save time by doing simple tests for allelism between your mutation and any other mutation that causes a similar phenotype. This type of testing is done by mating a homozygous or known heterozygous mouse, with a known, characterized, and mapped mutation, to a homozygous or known heterozygous mouse carrying the new mutation. If a mating between these animals produces offspring in the first generation with the same phenotype as both parents, then it can be interpreted that the new mutation is a remutation or allele of the known mutation. This type of test is simple if the mutation is recessive, but becomes much more complicated in a dominant mutation. This approach is described in more detail in Chapter 2, Determining the Genetic Basis of a New Trait.

Setting up matings for interspecific and intersubspecific crosses, needed for gene mapping (Chapter 2), comes with its own set of problems in colony management.

*Mus musculus castaneus* (CAST/EiJ) mice are much more aggressive than most common inbred strains of mice and are known to kill mice smaller than themselves if the mice are introduced to each other much after weaning age (approximately 3 weeks of age). This is true of both males and females. CAST/EiJ mice are also slow to reach sexual maturity. A mating between two CAST/EiJ mice or a CAST/EiJ male and a female of another strain may not produce their first litter until the mated pair has been together for as long as three months. It is important not to deem a mating "nonproductive" until at least this amount of time has passed. Female CAST/EiJ mice mated to older males of another strain may have their first litter sooner than three months, but again it is important not to discontinue the mating too soon. An increased success rate in starting a colony of mutant X CAST/EiJ crosses can be achieved by using CAST/EiJ females mated to mutant or known heterozygous males rather than vice versa (D. Boggess, personal observations).

## VII. CONCLUSIONS

In summary, identifying new mouse mutations and establishing colonies for these mutations requires careful observation, methodical work, and good record keeping on the part of both the technical and research staff in your facility. This is true whether the mutation is induced or naturally occurring. It will take a good bit of time and energy to reach the point where you will have a stable, reproducible colony carrying a mutation, and most if not all of the techniques mentioned in this chapter will probably be utilized. The chapters following this will provide details on how to further evaluate a new mutation, using breeding data to determine the genetics behind the mutation and how to best collect materials for the biological characterization of the phenotype.

## ACKNOWLEDGMENTS

This work was supported by grants from the National Institutes of Health (RR8911, AR43801, and CA34196). The authors thank K. Silva for critical review of the manuscript.

## REFERENCES

1. Lindsey, R.J., Boorman, G.A., Collins, M.J., Hsu, C-K., Van Hoosier, G.L. Jr., and Wagner, J.E., *Infectious Diseases of Mice and Rats*, National Academy Press, Washington, D.C., 1991.
2. Lindsey, R.J., Boorman, G.A., Collins, M.J., Hsu, C-K., Van Hoosier, G.L. Jr., and Wagner, J.E., *Companion Guide to Infectious Diseases of Mice and Rats*, National Academy Press, Washington, D.C., 1991.
3. Frith, C.H. and Ward, J.M., *Color Atlas of Neoplastic and Non-neoplastic Lesions in Aging Mice*, Elsevier, Amsterdam, 1988.
4. Mohr, U., Dungworth, D.L., Capen, C.C., Carlton, W.W., Sundberg, J.P., and Ward, J., *Pathobiology of the Aging Mouse*, ILSI Press, Washington, D.C., 1996.

5. Hebert, J.M., Rosenquist, T., Gotz, J., and Martin, G.R., FGF5 as a regulator of the hair growth cycle: evidence from targeted and spontaneous mutations, *Cell*, 78, 1017, 1994.

6. Sundberg, J.P., France, M., Boggess, D., Sundberg, B.A., Jenson, A.B., Beamer, W.G., and Shultz, L.D., Development and progression of psoriasiform dermatitis and systemic lesions in the flaky skin (*fsn*) mouse mutant, *Pathobiol.*, 65, 271, 1997.

7. Sundberg, J.P., Rourk, M.H., Boggess, D., Hogan, M.E., Sundberg, B.A., King, L.E. Jr., Sweet, H.O., Johnson, K., and Davisson, M.T., Juvenile alopecia (*jal*): a new mouse mutation with curly hair and focal alopecia on chromosome 13, *Exp. Dermatol.*, submitted.

8. Sundberg, J.P., Boggess, D., Hogan, M.E., Sundberg, B.A., Rourk, M.H., Harris, B., Johnson, K., Dunstan, R.W., and Davisson, M.T., Harlequin ichthyosis (*ichq*): a juvenile lethal mouse mutation with ichthyosiform dermatitis, *Am. J. Pathol.*, 151, 293, 1997.

9. Sundberg, J.P., The matted (*ma*) mutation, chromosome 3. *Handbook of Mouse Mutations with Skin and Hair Abnormalities: Animal Models and Biomedical Tools*, Sundberg, J.P., Ed., CRC Press, Boca Raton, FL, 1994, 345.

10. HogenEsch, H., Gijbels, M.J.J., Offerman, E., van Hooft, J., van Bekkum, D. W., and Zurcher, C., A spontaneous mutation characterized by chronic proliferative dermatitis in C57BL mice, *Am. J. Pathol.*, 143, 972, 1993.

11. Sundberg, J.P., The flaky tail (*ft*) mutation, chromosome 3. *Handbook of Mouse Mutations with Skin and Hair Abnormalities: Animal Models and Biomedical Tools*, Sundberg, J.P., Ed., CRC Press, Boca Raton, FL, 1994, 269.

12. Sundberg, J.P. and Sweet, H.O., The lethal milk (*lm*) mutation, Chromosome 2, in *Handbook of Mouse Mutations with Skin and Hair Abnormalities: Animal Models and Biomedical Tools*, Sundberg, J.P., Ed., CRC Press, Boca Raton, FL, 1994, 337.

13. Carlier, M., Roubertoux, P., and Cohen-Salmon, C.H., Differences in patterns of pup care in *Mus musculus domesticus*—Comparisons of eight inbred strains, *Behavioral and Neural Biology*, 35, 205, 1982.

14. Knight, C.H., Maltz, E., and Docherty, A.H., Milk yield and composition in mice: Effects of litter size and lactation number, *Comp. Biochem. Physiol*, 84A, 127, 1986.

15. Ragueneau, S., Early development in mice. IV. Quantity and gross composition of milk in five inbred strains, *Physiology & Behavior*, 40, 431, 1987.

16. Hanrahan, J.P. and Eisen, E.J., A lactation curve for mice, *Laboratory Animal Science*, 20, 101, 1970.

17. Luckey T., *Germfree Life and Gnotobiology*, Academic Press, New York, p. 495, 1963.

18. Pleasants, J.R., Rearing germfree cesarean-born rats, mice and rabbits through weaning, *Ann. N. Y. Acad. Sci.*, 78, 116, 1959.

19. Silver, L.M., *Mouse Genetics: Concepts and Applications*, Oxford University Press, New York, NY, 1995.

20. Snell, G.D. and Bunker, H.P., Histocompatibility genes of mice. V. Five new histocompatibility loci identified by congenic resistant lines on a C57BL/10 background, *Transplantation*, 3, 235, 1965.

21. McCune, J.M., Namikawa, R., Kaneshima, H., Shultz, L.D., Lieberman, M., and Weissman, I.L. The SCID-hu mouse: Murine model for the analysis of human hematolymphoid differentiation and function, *Science*, 241, 1632, 1988.

22. Walzer, P.D., Kim, C.K., Linke, M.J., Pogue, C.L., Wixson, S.K., Hall, E., and Shultz, L.D., Outbreaks of *Pneumocystis carinii* pneumonia in colonies of immunodeficient mice, *Infect. Immun.* 57, 62, 1989.

23. Shultz, L.D., Schweitzer, P.A., Hall, E.J., Sundberg, J.P., Taylor, S., and Walzer, P.D., *Pneumocystis carinii* pneumonia in *scid/scid* mice, *Cur. Topics Microbiol. Immunol.*, 152, 243, 1989.

24. Cunliffe-Beamer, T.L., personal experience.

25. Jones, E. and Krohn, P., Orthotopic ovarian transplantation in mice, *J. Endocrinol.*, 20, 135, 1960.

26. Robertson, G., Ovarian transplantation in the house mouse, *Proc. Soc. Exp. Biol. (NY)*, 44, 302, 1942.

27. Robertson, G., An analysis of the development of homozygous yellow mouse embryos, *J. Exp. Biol.*, 89, 197, 1942.

28. Russell, W. and Hurst, J., 1945. Pure strain mice born to hybrid mothers following ovarian transplantation, *Proc. Natl. Acad. Sci.*, 31, 267, 1945.

29. Snell, G.D. and Stimpfling, J.H., Genetics of tissue transplantation, in *Biology of the Laboratory Mouse*, 2nd ed., Green, E.L., Ed., Dover Publications, New York, 1975, 458.

30. Stevens, L., A modification of Robertson's technique of homoiotopic ovarian transplantation in mice, *Transplantation Bull.*, 4, 106, 1957.

31. Tanioka, Y., Tuskada M., and Esoki, K., A technique of ovarian transplantation in mice, *Exp. Anim.*, 22, 15, 1973.

32. Cunliffe-Beamer, T.L. Biomethodology and surgical techniques, in *The Mouse in Biomedical Research Vol. 3*, Foster, H.L., Small, J.D., and Fox, J.G., Eds. Academic Press, New York, 1983, 425.

# 2   Determining the Genetic Basis of a New Trait

*Xavier Montagutelli*

## CONTENTS

## I. INTRODUCTION

Finding a mouse with an abnormal phenotype is not a rare observation in an animal facility. In some instances, this can be the consequence of infectious diseases, abnormal environmental parameters, intoxication, or any other acquired condition. When this is the case, one may find other individuals with the same phenotype, but, unless the causative agent is properly identified, such phenotypic deviants will be of little use as animal models. In other instances, the abnormal phenotype has a genetic origin, be it the consequence of a mutation in a single gene or due to the effect of a particular allelic combination.

In order to make a useful model from the initial observation, it is important to determine the genetic nature of the phenotype, establish its mode of inheritance, and identify the causative gene(s). The aim of this chapter is to review the standard protocols currently used to perform these genetic studies.

## II. IS THE PHENOTYPIC DEVIANT GENETICALLY DETERMINED?

This is the first question to be addressed. The answer will come from two directions. First, one should investigate any change in "environmental" factors that could account for the phenotype: infectious diseases, food, temperature, humidity, breeding and husbandry procedures, etc. For example, with alopecia and skin lesions in general, one should always look first for ectoparasites or malnutrition. Notably, if the abnormal phenotype is due to an element of the environment, similar cases (with various degrees of severity) will likely be observed in the colony.

Second, to assess the genetic control of the phenotype, one should aim at producing more progeny from the same breeding pairs, to cross the affected mouse with unaffected littermates (when possible), and to cross unaffected littermates with each other or with either parent (often daughters with their father). If the trait is genetically determined, one of these crosses should yield new affected offspring, which will establish a segregating colony (see Chapter 1 on colony establishment).

In most cases, once the first phenodeviant has been identified, sister–brother matings should be the rule for maintaining the colony, since inbreeding increases the chance of producing individuals homozygous for a recessive mutation and homogenizes the genetic background. When the first phenodeviant appears on a mixed (outbred) background and inbreeding is being performed, it is frequent that the phenotype changes more or less profoundly with generations, along with the evolution of the genetic background. If the mutation occurs on a mixed background, repeated crosses with an already established inbred strain (called the *recipient strain*) should be made to "transfer" the mutation onto a standard genetically homogeneous background by the development of a congenic strain (see Chapter 1).

## III. IS THE GENETIC CONTROL SIMPLE (MONOGENIC TRAIT) OR COMPLEX (POLYGENIC TRAIT)?

Once it has been established that the trait is genetically determined (even though some environmental parameters may play a role), one should ask whether the trait is controlled by a single gene or by multiple genes and, in the first case, whether it is dominant or recessive, autosomal, or sex-linked. These questions can be answered from the results of a few simple crosses. When possible, it is desirable to perform two reciprocal crosses, where affected females/males are mated with unrelated, unaffected males/females, preferably from the same inbred strain, to produce 30 to 40 F1 progeny in each cross.

If 50% of the F1 progeny, in both sexes, are affected in the two crosses, it is likely that the trait is determined by a single, dominant gene carried by an autosome. In this case, intercrossing of affected F1 progeny should yield either 75% affected and 25% unaffected (dominant viable mutation), or 66% affected and 33% unaffected (dominant mutation with lethality if homozygous), or 50% affected, 25% unaffected, and 25% with a new abnormal phenotype, probably more severe than the original one (semidominant mutation).

A sex-linked dominant mutation is suggested by the observation that females show two types of abnormal phenotypes, whereas males show only one of these types (usually the more extensive). This is the consequence of X inactivation which results, in females, in a patchwork of cells expressing randomly either the normal or the mutated allele.[1] Linkage to the X chromosome is confirmed when the cross of an affected female with an unaffected male yields 50% affected F1 progeny in both sexes (with females showing the milder phenotype and males the more severe), whereas the reciprocal cross yields only unaffected males and affected females.

In the third case, the cross of an affected individual with an unaffected, unrelated partner will yield no affected F1 progeny. Intercrossing these progeny will yield F2 affected mice in a 25% proportion of the total, in both sexes. Crossing an unaffected F1 with its affected parent will yield 50% affected progeny, in both sexes. All these results are expected from a recessive mutation at a single autosomal locus.

Simple Mendelian traits, be they inherited in a recessive or dominant manner, can express with incomplete penetrance, which means that not all individuals carrying the mutant gene will show an abnormal phenotype. Incomplete penetrance can be detected in the crosses described above by a reduced yield of affected progeny. Penetrance can be profoundly influenced by the genetic background, hence the importance of performing these crosses using individuals of the same inbred strain.

The majority of phenotypic deviants that can arise in a mouse colony either spontaneously or after treatment with a mutagen will fall into one of these categories. Less frequently, the crosses mentioned above will give different proportions of phenotypes or will even yield new classes of phenotypes. In such cases, one should suspect a polygenic inheritance. Briefly, identifying the genes controlling a complex polygenic trait is performed through the establishment of a mapping cross with an unrelated strain that will identify regions of the genome associated with the trait. Narrowing these regions down to candidate genes requires the development of congenic strains. The study of complex traits has been thoroughly described by others[2] and is beyond the scope of this chapter.

## IV. IS THE NEW MUTATION ALLELIC TO A PREVIOUSLY DESCRIBED MUTATION?

Clinical and histological evaluation of a new mutation is sometimes reminiscent of other, previously characterized mouse mutations. In this case, and only if the new mutation is recessive, it is possible to prove or to disregard the hypothesis that this mutation is a new allele (or remutation) at the same locus as one of the known mutations. The principle is that mice heterozygous for both mutations will show an abnormal phenotype if the two mutations affect the same gene, whereas they will appear normal if they affect different genes.

Practically, either known heterozygous or homozygous mice of the new mutation are mated with either known heterozygous or homozygous mice of the known mutation. The observation that a fraction of the progeny show an abnormal phenotype close to that of either parental mutant demonstrates that the two mutations affect the same gene. If the two mutations are allelic, the mating of a mouse homozygous for one mutation with a mouse homozygous for the other should yield only affected progeny. If one of the two parents is homozygous for one mutation and the other is heterozygous for the other, an equal proportion of affected and phenotypically normal progeny should be obtained. Finally, if the parents are heterozygous for different mutations that are allelic, one littermate out of four should be affected.

In the case of semidominant or dominant mutations, proving allelism is much more difficult to achieve and will be based on (1) identical map position and (2) either the occurrence of a new phenotype, or the frequency of the different classes of phenotypes, when crossing heterozygous mice.

## V. GENETIC MAPPING

### A. INTRODUCTION

There are at least four reasons for localizing a new mutation on the genetic map of the mouse. First and as mentioned above, the position of the mutation will be instrumental in evaluating its possible allelic relationships with other mutations suggested by the phenotype. Second, genetic mapping is the first step toward the identification of the mutated gene. This is particularly important with the exponentially increasing number of DNA sequences that are being cloned and positioned on the genetic map of the mouse, in particular expressed sequence tags referred to as ESTs,[3] which provide useful hints for the discovery of the mutated gene. The third reason is based on the well-known homologies that exist between the mouse and the human chromosomes, such that three out of four new genes that are being mapped in one species can be assigned provisionally to a chromosomal region in the other without any experimental work.[4] In the case of a new mouse mutation that mimics a human genetic disease, knowing where the mutation maps in the mouse will suggest a possible location for the gene responsible for the human disease. Lastly, the genetic mapping of the mouse mutation will provide genetically linked polymorphic markers that will be very useful, for example, to identify affected mice before the symptoms develop or to distinguish heterozygous from wild-type mice in the case of a recessive mutation. These arguments strongly support the mapping of all new mutations as one of the very first experiments to be done.

While a newly cloned DNA fragment can be localized very precisely on the genetic map with the help of already established and characterized mapping panels (such as The Jackson Laboratory interspecific mapping panels,[5] somatic hybrids, or radiation hybrid panels),[6] mapping a new mutation requires that a cross be set up in which the mutation segregates together with a large number of polymorphic genetic markers. The two most common breeding schemes start with the production of F1 hybrids between two strains (one of which carries the mutation); these F1 progeny can either be intercrossed to produce an F2 generation, or they can be crossed with mice of either of the two parental strains (depending on the mode of inheritance of the mutation) to produce a backcross (BC) generation.

The accuracy of a mapping cross is limited by the number of progeny generated and genotyped and by the availability of polymorphic markers in the region that contains the mutation.

### B. HOW TO CHOOSE THE PARENTAL STRAINS

One of the parental strains to be used in the mapping cross must be the colony where the mutation is being maintained. In most instances, this colony will be derived from a laboratory inbred strain or outbred colony. An inbred genetic background is advantageous because it simplifies the linkage analysis.

The mapping cross can be set up using another laboratory inbred strain. Such strains exist in large numbers, and they usually breed well when crossed with others. The F1 progeny produced show increased fertility and will easily produce large numbers of F2 or BC progeny. It is recommended to choose as a partner a laboratory strain that is as unrelated as possible to the strain in which the mutation arose, with the aim of increasing the degree of polymorphism between the two parental strains. However, from the history of the common mouse inbred strains,[7] we learn that most of them are derived from a limited subset of progenitor mice with origins mostly in the *Mus musculus domesticus* but also in the *Mus musculus musculus* subspecies (for this reason such crosses are called intraspecific crosses) and therefore show a minimal level of interstrain polymorphism (they will be thereafter referred to as *Mus m. musculus/domesticus*). For example, inbred strains C57BL/6 and DBA/2, two of the most genetically different inbred strains, carry different alleles in only 35-40% of the microsatellite markers known to date.[8] This drawback is acceptable if the aim of the cross is to map the mutation with a precision of 1 to 2 cM (centiMorgan), but it is a serious limitation to consider before starting a high-resolution mapping project. In particular, such intraspecific crosses are not recommended if positional cloning of the mutation is to be undertaken.

For this reason, crosses between inbred strains of mice derived from different species have become widely used for mapping purposes. *Mus spretus* is the most distant species from *Mus m. musculus/domesticus* that still allow the production of fertile F1 hybrids.[9] It offers the highest level of polymorphism when crossed with laboratory inbred strains.[10] However, such interspecific crosses present two major drawbacks that can hamper a mapping project. First, they usually yield a small number of progeny. Second, F1 males are always sterile, which rules out the possibility of producing an F2 generation. An additional, but variable, difficulty comes from the possible influence of the highly divergent genetic background on the expression of the mutation.

To circumvent these problems, a number of investigators are now using inter-subspecific crosses that involve inbred strains from the *M. musculus musculus* or *M. musculus castaneus* subspecies. Such crosses offer a level of polymorphism that is almost as high as with interspecific crosses[11] but with much easier breeding and with the ability to produce intercrosses.

## C. How to Choose the Breeding Scheme

The breeding scheme, i.e., the genotypes of the parents to be mated, depends on the mode of inheritance of the mutation. The possibilities are illustrated in Figure 2.1 (dominant mutation) and Figure 2.2 (recessive mutation). It should be noted that, in several instances with recessive mutations, it is necessary to produce four times more progeny than will be useful for the mapping because it is not always possible to deduce the genotype at the disease locus based on the phenotype of the mice (phenotypically normal mice can be either +/+ or +/*m*).

### Dominant mutation: backcross

OUTCROSS

M. m. castaneus          M/+

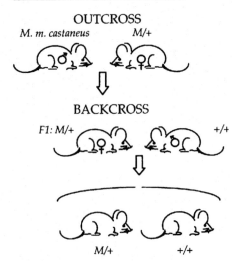

BACKCROSS

F1: M/+                                              +/+

M/+                    +/+

**FIGURE 2.1** Breeding scheme for the mapping of a dominant mutation. Heterozygous affected females are outcrossed with *Mus musculus castaneus* (CAST/EiJ) males. Affected F1s (males or females) are backcrossed with unaffected mice from the colony carrying the mutation. Genotyping is performed on both affected and unaffected backcross progeny.

Whereas dominant mutations are mapped in a backcross (see Figure 2.1), recessive mutations can be mapped using either a backcross or an intercross (see Figure 2.2A through D). It is worth discussing the advantages and limitations of these two schemes.

In a backcross (BC), the only meiosis in which informative recombination occurs is in the F1 parent. For this reason, the analysis is very straightforward because all chromosomal breakpoints can be directly deduced from the genotypes of the progeny. Gene ordering can be performed by minimizing the number of double recombinants without the help of specialized computer software, even though such programs make it easier. Another advantage comes from the fact that it is possible to produce a large number of BC progeny by mating a few heterozygous F1 males with a large number of females, whose genotype depends on the nature of the mutation. Finally, since the BC progeny have 75% of their genome coming from the inbred strain where the mutation has been characterized, there will be less phenotypic variation due to the influence of modifier genes coming from the alien strain than in the case of intercross (F2) progeny. The only disadvantage of a backcross is that only one meiosis is analyzed per mouse and, compared with an F2 cross, twice as many progeny have to be genotyped in order to get the same accuracy of the map position.

Intercrosses are particularly useful with recessive mutations maintained in a small colony. In this case, a small number of males from the colony can be used to

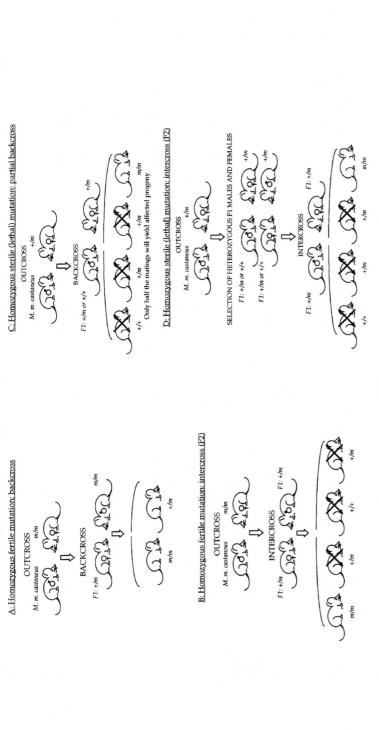

**FIGURE 2.2** Breeding schemes for the mapping of a recessive mutation. A and B apply to mutations where homozygous affected mice are fertile. C and D apply to mutations where only heterozygous carrier mice are fertile. (A) backcross: Homozygous affected females are outcrossed with *Mus musculus castaneus* (CAST/EiJ) males. Heterozygous F1s (males or females) are backcrossed with homozygous affected mice. Genotyping is performed on both affected and unaffected backcross progeny. (B) intercross: Homozygous affected females are outcrossed with *Mus musculus castaneus* males. Heterozygous F1s are intercrossed. All matings yield a quarter of affected F2 progeny which are genotyped. (C) partial backcross (or backcross–intercross): Heterozygous females are outcrossed with *Mus musculus castaneus* males. F1s (males or females) are genotyped. Since only half the F1s are heterozygous for the mutation, only half the crosses yield a quarter of affected F2 progeny which are genotyped. (D) intercross: Heterozygous females are outcrossed with *Mus musculus castaneus* males. F1 males and females are first mated with heterozygous mice from the colony to identify those F1s that are heterozygous for the mutation. These are then intercrossed. All matings yield a quarter of affected F2 progeny which are genotyped.

produce a number of F1 mice that will then generate a large F2 generation. Compared to a backcross, each progeny yields twice as much genetic information because it carries an F1 chromosome from each parent. This is of special interest for high-resolution mapping projects. However, in the case of a recessive, homozygous fertile mutation, a backcross (Figure 2.2A) is generally preferred because an intercross will make use of only a quarter of the progeny (Figure 2.2B). The data analysis is not as simple as that of a backcross, and it is usually achieved with the help of computer software.

## D. How Many Progeny Should Be Produced? How Many Markers Should Be Typed?

These are the next questions to be addressed once a breeding scheme has been selected, and they are related. The number of progeny determines the ability of the cross to resolve closely linked loci as well as its power to detect linkage between distant markers. If M is the number of informative meioses analyzed and N the number of progeny (M = N for a backcross, and M = 2N for an intercross), the smallest distance that the cross can estimate is 1/M.

Also dependent on the size of the cross, the "swept radius" is defined as the length of a chromosome interval on either side of a marker locus within which linkage can be detected with a certain level of significance. If $r$ is the fraction of chromosomes that have recombined between the two loci, the departure of $r$ from 50% (expected under the null hypothesis of no linkage) can be tested by the chi-square with one degree of freedom:

$$\chi^2 = \frac{[M(1-r)-Mr]^2}{M} = M(1-2r)^2$$

One can choose the number of progeny to be produced based on the number of marker loci that one can genotype; on the contrary, one can estimate the number of marker loci to be genotyped based on the number of progeny that will be produced. Now that large numbers of polymorphic markers are available, the limiting factor is often the number of progeny that can be produced. Let us take an example. At the significance level of 0.001, $\chi^2 = 10.8$. With a backcross of 70 mice (M=70), it is possible to detect linkage at this significance level between marker loci separated with a recombination fraction of 0.3 (which corresponds approximately to a distance of 35 cM, see below). The same power would be obtained with an F2 of 35 informative progeny. In the case of mapping a new mutation inherited as a single gene Mendelian trait, it means that, with a cross of this size, genotyping the progeny for one marker every 70 cM is sufficient to find the chromosomal segment where the causative gene maps. The most proximal and distal markers should be chosen at a distance of 35 cM from either chromosome ends. Hence, less than 50 genetic markers are needed to scan the genome completely, provided they are evenly spaced along the genome map.

Once the mutation has been mapped to a chromosomal segment, additional genetic marker loci of this region have to be genotyped to define the location of the gene more accurately. The ultimate goal for fine mapping achievable with the cross will be to find genetic markers that are located on both sides of the mutant locus, with only one chromosome recombining between the mutant locus and either marker.

Finally, the number of progeny to be produced depends on the ultimate goal of the mapping project. If the aim is to produce a map with moderate resolution acceptable for publication, 60 to 80 recombinant chromosomes should be obtained (60 to 80 BC, 30 to 40 F2). If the aim is to develop a high-resolution map as the starting point toward the positional cloning of the mutant locus, it is necessary to analyze 600 to 1500 chromosomes. However, only 50 to 100 chromosomes will be analyzed in the first step to assign the mutant locus to a chromosomal segment.

## E. Phenotyping the Progeny

Once the type and the size of the cross has been decided, the appropriate number of matings should be set up to produce the progeny. It should be remembered that, with interspecific or intersubspecific crosses, not all matings will be fertile and assuming that 10 to 20% of them will not yield progeny provides a reasonable safety margin. In the case of interspecific backcrosses, only F1 females will be useful to generate the second generation because all F1 males are sterile.

Each BC or informative F2 mouse produced will be subjected to phenotype assessment, and tissues (such as spleen, kidney, liver, or brain) will be collected to prepare good quality DNA for the genotyping of molecular markers. One should also collect DNA from every parent and keep detailed breeding records, in case unexpected genotypes or alleles are recovered in the test progeny. This is absolutely necessary if the mutation arose on a mixed genetic background. Because individual mice in the colony where the mutation is maintained may carry different alleles at several marker loci, unexpected genotyping data can be explained with pedigree analysis, provided appropriate material was collected during the establishment of the cross.

Special emphasis should be placed on the importance of phenotyping the progeny very carefully. Accurate and thorough descriptions of phenotypes should be recorded for each mouse. Some mutations have such an obvious phenotype that it is unlikely that misclassification can occur. However, many mutations show more subtle features and great care should be taken to properly determine the phenotype of each progeny. In addition, interspecific or intersubspecific crosses result in the mixing of two evolutionary divergent genomes, and it is not a rare observation that affected F2 or BC progeny show a markedly different phenotype compared to that of affected mice from the inbred colony where the mutation is maintained. This is due to interactions between the mutation itself and various numbers of modifying genes.[12]

Therefore, it is advisable to combine different criteria to assess the phenotype, such as clinical examination, biopsies for histology or immunohistochemistry, blood samples for biochemistry, etc., depending on the mutation. Incorrect phenotyping can result in aberrant double crossovers involving the mutant locus and the two closely flanking loci. As a consequence, it is not possible to ascertain the position of the mutant locus.

When a mutation is incompletely penetrant, only a fraction of the mice with the mutant genotype express an abnormal phenotype. In this case, only mice which are obviously affected (the manifesting class) should be genotyped for marker loci because they are the only ones for which genotype at the mutant locus can be established unambiguously.

## F. THE GENETIC MARKERS

Several different types of genetic markers can be used to determine linkage. Molecular markers are codominantly inherited and can be typed easily. The most popular ones, those commonly used in the assignment of a new gene to a particular chromosomal region, are considered below. However, there are other types of markers that can be useful, including biochemical (also codominant) or immunological markers and a variety of molecular markers that take advantage of moderately or highly repeated DNA sequences.[13,14] These markers are often used in the high-resolution phase of a mapping project, where every available marker is useful to saturate the genetic map.

The genetic map of each chromosome of the mouse is constantly updated from data published in the literature by a committee that publishes a yearly report containing a consensus map in a special issue of the journal *Mammalian Genome*. These reports, as well as all information concerning genetic markers in the mouse, can be retrieved on-line in the *Mouse Genome Database*, maintained at The Jackson Laboratory and accessible via the Internet at http://www.informatics.jax.org/MGD (see Chapter 13).

## 1. Microsatellites

The human and mouse genomes contain large numbers of tandem repeats of a small motif of 1 to 4 nucleotides, called *microsatellites*, mostly located in noncoding sequences.[15] The most common motif, a repeat of CA, represents more than $10^5$ copies.[16] The number of repetitions of the motif at each chromosome site is consistent within an inbred strain but is variable between strains so that the motif can be used as a molecular marker known as an SSLP, simple sequence length polymorphism.[17] By designing PCR primers specific for the unique flanking sequences of a particular repeat, one can amplify a DNA segment, the length of which depends on the number of CAs of the repeat, a characteristic of a particular inbred strain. Over 6000 such microsatellite sequences have been cloned in the mouse by the team headed by Eric Lander at MIT,[8] so-called "MIT markers." These microsatellites span the entire genome, with the exception of the Y chromosome. Primers have been designed and the size of the amplified fragments determined for each microsatellite for a panel of a dozen inbred strains, including CAST/Ei (*Mus musculus castaneus*), and several laboratory strains, such as C57BL/6J, C3H/HeJ, or BALB/cJ. These primers are commercially available from Research Genetics (Huntsville, AL), and all sequences, polymorphism data, and map positions are available via the Internet (http://www.genome.wi.mit.edu/cgi-bin/mouse/index). For these reasons, the use of microsatellites as genetic markers is strongly encouraged for the mapping of a new mutation.

Microsatellites have rapidly become the markers of choice in a number of species (including mice, humans, and a number of domestic animals) because they are easy to characterize. The typing involves a PCR reaction with the genomic DNA of each progeny as template. In most cases, PCR can be performed using nonradioactive nucleotides and standard procedures, such as those recommended by the manufacturer of the *Taq* polymerase enzyme. The PCR products, which are usually 80 to 220 bp in length, are separated on either a horizontal 4% agarose gel stained with ethidium bromide (use special agarose such as Nu-Sieve or Metaphor from FMC Bioproducts, Rockland, ME) or a vertical nondenaturing 8% polyacrylamide gel stained with either ethidium bromide or silver nitrate. The type of gel depends on the expected size difference of the PCR products between the two parental strains: agarose gels can resolve size differences of 4 to 5 bp in a 100-bp fragment, whereas sequencing polyacrylamide gels can identify a single base pair difference in a 250-bp fragment. Some researchers incorporate radioactive nucleotides in the PCR reaction and separate the products on a nondenaturing polyacrylamide gel. After migration, the gel is dried and an autoradiograph prepared.

An even more rapid method of genotyping progeny from a linkage cross is to pool DNA from 20 mutant mice and screen the pool for multiple markers simultaneously.[18]

When using microsatellites with interspecific or intersubspecific crosses, one should be aware that PCR primers have been designed according to DNA sequences of laboratory strains (*Mus musculus domesticus*) and that the corresponding sequences in *Mus musculus castaneus* or *Mus spretus* might be slightly different. When PCR is performed on a DNA sample from a heterozygous mouse, which contains both types of sequences, there is sometimes preferential amplification of the *Mus musculus domesticus* allele, resulting in mistyping.[19] Coamplification of both alleles should therefore always be checked.

Finally, it should be noted that most microsatellites are located within noncoding or even intergenic sequences and correspond to anonymous markers.

## 2. Restriction Fragment Length Polymorphisms

Restriction fragment length polymorphisms (RFLPs) have been used for over 15 years as molecular markers. They have allowed the development of the first high-density, multilocus maps before the advent of microsatellites.[20, 21] RFLPs are based on the presence of polymorphic sites for restriction enzymes at the DNA level. There are two methods for using such sites as genetic markers. The classical method uses the Southern blotting technique and hybridization with a radiolabeled DNA probe that is complementary to a DNA fragment flanked by the polymorphic restriction site. The position of the band revealed by autoradiography depends on the length of the fragment, i.e., of the position of the nearest restriction sites of the enzyme used to digest the genomic DNA.

A more recent technique combines PCR amplification of a DNA fragment encompassing the polymorphic restriction site, followed by the digestion of PCR products with the restriction enzyme. If the restriction site is present, the PCR product is cut and appears as two bands on an agarose gel stained with ethidium bromide, whereas it appears as a larger, single band if the restriction site is absent. This technique is much cheaper and uses much less DNA than the former but is less general in that it requires that 100 bp of genomic DNA flanking the polymorphic restriction site be known.

RFLP markers have two advantages: (1) they are often associated with genes and (2) they do not require that the sequence be known. They are very useful to investigate the linkage relationships between the mutant locus and candidate genes that are suggested by both map position and function. However, because they require more work and material, and because it is often troublesome to obtain an adequate number of probes, they are no longer used to scan the genome in order to identify the chromosomal segment where the new mutation maps.

## 3. Single-Strand Conformation Polymorphisms

Single-strand conformation polymorphisms (SSCPs) have been proposed more recently as a simpler and more abundant source of polymorphism than RFLPs.[22] Briefly, a short sequence of 100 to 250 bp is amplified by PCR using radiolabeled nucleotides. The PCR product is then diluted and denatured as single-stranded DNA molecules before being loaded on a nondenaturing polyacrylamide gel. Single-stranded DNA molecules fold to acquire the three-dimensional conformations with the lowest level of energy, which are tightly dependent on their sequence. Two DNA segments of 100 bp differing in a single nucleotide will often adopt different conformations and, hence, will migrate at a different speed in the gel. After migration, the gel is dried and an autoradiograph prepared.

This technique is very powerful in that it can detect almost any nucleotide change. It is an interesting alternative to RFLPs for the evaluation of candidate genes. It is applicable to virtually any DNA segment, provided one or two hundred base pairs of sequence are known.

## G. Linkage Analysis and Assignment of the Mutation to a Chromosomal Interval

For the mapping of a mutation inherited as a monogenic Mendelian trait, one marker should be genotyped at a time, immediately followed by linkage analysis, to avoid unnecessary genotyping work. It is not required that the entire genome be scanned, as soon as linkage has been established between the mutation and at least one genetic marker with a high degree of confidence. Genetic markers can be analyzed in whatever order is convenient. In some instances, the breeding records can suggest or exclude certain map positions, for example, in the case where it has been observed that the mutation cosegregates with a coat-color marker such as agouti (Chr 2) or albino (Chr 7).

Even when a large number of progeny are available (for example, when the ultimate goal of the mapping project is to develop a high-resolution map with over 1000 chromosomes analyzed), this first step will be performed on a small subset of 50 to 100 chromosomes.

Although linkage analysis could be performed by hand, it is recommended that the phenotyping and genotyping data be stored in a specialized computer program such as the Apple Macintosh-based Map Manager program developed by Ken Manly,[23] the Mapmaker program developed by Eric Lander,[24] or the Gene-Link program available at http://bioweb.pasteur.fr/seqanal/soft-pasteur.html. These data essentially consist of a large matrix with each row representing a locus, each column representing a mouse, and each cell containing a genotype. In the case of a dominant mutation (Figure 2.1), and in the case of a backcross with a recessive fertile mutation (Figure 2.2A), affected mice are denoted as homozygous for the *Mus m. musculus/domesticus* allele at the mutant locus, and unaffected mice are denoted as heterozygous. In the case where only affected progeny can be analyzed (Figure 2.2B, C, and D), all mice are denoted as homozygous for the *Mus m. musculus/domesticus* allele at the mutant locus.

Linkage evaluation between the mutant locus and a marker locus is accomplished by comparing the distributions of genotypes for these two rows. Under the null hypothesis of no-linkage, alleles reassort at random. In the cases illustrated in Figure 2.1 and Figure 2.2A, affected mice will be either heterozygous or homozygous for the *Mus m. musculus/domesticus* allele at the marker locus in equal numbers, as will the unaffected progeny (see Figure 2.3A). The observed distribution is compared to this model using a simple chi-square, and linkage is concluded when a significant departure from the null hypothesis is observed. For a backcross, the recombination fraction between the mutant and the marker loci is easily calculated from the number of mice with dissimilar genotypes divided by the number of mice genotyped for both markers, as illustrated in Figure 2.3B. For an intercross, it is necessary to count each chromosome and, again, this is more easily accomplished with the help of a computer program.

Genetic distances are expressed in centiMorgans (cM). One centiMorgan corresponds to a recombination fraction of 1%. While centiMorgans are additive distance units, recombination fractions are not, under the assumption that crossovers occur independent of one another. For example, under this assumption, a 30-cM interval should show a recombination fraction of 30% diminished by $0.3 \times 0.3 = 9\%$ to account for undetected double recombinant, increased by $(0.3)^3 \approx 3\%$ to account for triple recombinants, so that the apparent recombination fraction will be $\approx 23.5\%$. However, experimental data have shown for a long time that when a crossover occurs at some point along the chromosome, other crossovers occur less frequently than expected in its vicinity. This phenomenon, known as *interference*, has led several authors to propose the so-called "mapping functions" which convert recombination fractions into cM distances under different models of interference. In the mouse, interference is so strong that, for recombination fractions up to 20%, double recombinations are very rare, and cM distances equal recombination percents.

At some point, a marker locus will be found that shows significant linkage with the mutant locus. The next step should be to analyze the next proximal and distal markers in order to assign the mutation to a precise chromosomal segment. If the marker loci are 30 cM or more apart, linkage will be observed most likely with one or even with two adjacent loci. The correct order of these loci is the one that implies the minimal number of double recombination events. Linkage programs offer more or less automated ordering of loci.

It is possible that, during the course of mapping, weak linkage is found between the mutant and a marker locus. If no linkage is found with either marker flanking the marker locus, it will be concluded that the "linkage" detected was due to sampling fluctuations and that genotyping should be resumed.

**A - Absence of linkage between the mutation and a genetic marker**

(dominant mutation)      *d: Mus m. domesticus allele*
                                    *c: Mus m. castaneus allele*

|  | **Affected** | | **Unaffected** | | |
|---|---|---|---|---|---|
|  | *d/d = 48* | | *d/c = 48* | | |
| Marker X | $d/d^{(1)}$ | $d/c^{(2)}$ | $d/d^{(2)}$ | $d/c^{(1)}$ | |
| Observed | 27 | 21 | 20 | 24 | Chi-square = 1.1 |
| Expected if no linkage | 24 | 24 | 22 | 22 | P > 0.05 |

(1) non-recombinant class of genotypes
(2) recombinant class of genotypes

**B - Genetic linkage between the mutation and a genetic marker**

(dominant mutation)      *d: Mus m. domesticus allele*
                                    *c: Mus m. castaneus allele*

|  | **Affected** | | **Unaffected** | | |
|---|---|---|---|---|---|
|  | *d/d = 48* | | *d/c = 48* | | |
| Marker X | $d/d^{(1)}$ | $d/c^{(2)}$ | $d/d^{(2)}$ | $d/c^{(1)}$ | |
| Observed | 32 | 16 | 15 | 29 | Chi-square = 9.8 |
| Expected if no linkage | 24 | 24 | 22 | 22 | P = 0.002 |

(1) non-recombinant class of genotypes      Recombination fraction = 31/92 = 0.337
(2) recombinant class of genotypes

**FIGURE 2.3** Genetic linkage analysis in a backcross (case of a dominant mutation). Each backcrossed (BC) progeny is classified into one of four classes, according to its phenotype and its the genotype at the marker under test. The distribution of the BC mice in the four classes is compared to that expected under the hypothesis of no genetic linkage between the mutation and the marker locus, using a chi-square. (A): If the marker X is not linked to the mutation, the observed numbers are approximately equal in the four classes. (B): If the marker X is linked to the mutation, there is a significant excess of mice in the two classes of nonrecombinant genotypes (denoted as 1). The recombination fraction is calculated from the number of recombinant mice over the total number of progeny genotyped for marker X.

If some marker loci are too distant from each other or from either chromosome end, it may be necessary to analyze additional markers in order to find linkage with a high level of confidence.

## H. REFINING THE MAP POSITION

After the mutant locus has been assigned to a chromosomal region of 20 to 30 cM, it is necessary to type additional markers in this interval in order to refine the map position, in particular relative to known genes that could be candidates for the mutant locus. It is recommended that the chromosomal region be divided into smaller intervals of 5 cM by choosing more microsatellite markers. At this stage, the genotyping will still be performed on the same limited set of animals (50 to 100). Because genetic distances between these markers are very small, double crossovers are very unlikely and, if one is found, it should be carefully checked. It is possible that a marker does not recombine with the mutation in any of the mice. The aim should be to find a pair of markers, located on both sides of the mutation, which do show recombination with the mutation in at least one animal (and preferably in just one for each locus). At this point, the mapping data are strong enough to be included in a publication.

## I. HIGH-RESOLUTION MAPPING

If the aim of the mapping project is to clone the mutant gene, it is necessary to breed a large number of mice (in order to analyze several hundreds to 1000 or more chromosomes). Again, depending on the breeding scheme, it may be necessary to produce four times more mice than will be genotyped. All the mice retained for analysis will first be genotyped for the two closest genetic markers recombining with the mutation. All mice that do not recombine between these markers will be discarded because they are of no help in reducing the size of the interval. If the genetic interval is 1 cM, only 1% of the mice produced will be considered further. These rare recombinant mice are of high value and care should be taken to confirm their genotype and to collect enough material from them. By genotyping all markers available in the chromosomal region on these recombinant mice, one should be able to reduce even more the candidate genetic interval and to obtain an unambiguous order for all markers that show recombination with each other and/or with the mutation. Nonmutant recombinant progeny should be kept alive and test mated with mutation carriers to determine whether they are +/+ or +/m.

## J. TESTING CANDIDATE GENES

The markers that have been used for the mapping of the mutation were anonymous markers, most of which are not related to genes. When a small genetic interval has been defined where the mutant locus must lie (not necessarily with hundreds of mice), it is desirable to look for known genes already mapped to the same chromosomal region that could be candidates for the mutation on the basis of both map position and function.

Testing a gene as a candidate can require a lot of effort, and it is desirable to consider only those that reside within the defined genetic interval. Comprehensive chromosome maps, such as those established by the Chromosome Committees and published yearly in *Mammalian Genome* or available through the *Mouse Genome Database,* are useful in identifying all such genes. However, these "consensus" maps result from the data published in the literature, which are not always informative enough to establish gene order with a high degree of confidence. It is therefore necessary to refer to the primary data from which these maps were constructed in order to make sure which genes can be excluded from the genetic interval and which ones should be considered. Gene order can never be ascertained from genetic distances but only from the observation of recombination events within the same cross. The only true test of whether the gene lies within the mutation's interval is to type it in recombinant progeny from the linkage cross.

Not only mouse genes with known function should be studied as candidate genes. Human cloned and mapped genes should also be considered, taking advantage of the chromosomal homologies between mice and men. Moreover, large efforts are being made in the human toward the cloning of hundreds of thousands of small DNA fragments which correspond to partial sequences of most genes expressed in a particular tissue.[25] These ESTs will be placed on the human genetic map, so that they will provide even more position-based candidates for a particular mutation. One should keep in mind that most of these sequences are partial and that neither the complete coding sequence nor the function of the corresponding gene is known. However, ESTs provide very useful molecular tools and make the complete cloning of the gene much easier.

The first step in testing a candidate gene is to use it as a genetic marker in the mapping cross and to look for recombination between this gene and the mutation. When a DNA probe is available, the RFLP method can be tried with a number of restriction enzymes in order to find polymorphism. If DNA sequences are available from databases, primers can be designed either to generate a specific probe for hybridization or for SSCP analysis. Only those genes that do not show recombination with the mutation should be selected for further analysis. If the minimal genetic interval has been established on the basis of a large number of mice, most candidate genes will be discarded during this first step, and few will be submitted to sequence analysis.

The next step will comprise a variety of techniques, depending on the size of the gene, its genomic structure (when it is known), and its expression. For example, if a tissue is known where the gene is expressed, Northern blotting analysis can be performed, and one will look for reduced level of expression or for one or more band(s) with abnormal size. RT–PCR followed by sequencing can be used to identify mutations in the coding sequence. If the genomic structure is known, primers can be designed to PCR-amplify and sequence all exons and exon/intron boundaries.

A number of mouse mutations have already been cloned by this candidate gene approach (without the need for constructing a physical map), and the rapid development of ESTs will provide many new hints.

## K. POSITIONAL CLONING OF THE MUTATION

If no candidate gene has been found, or if all of them have been discarded, the cloning of the mutation requires constructing a physical map of the chromosomal region and looking for all possible expressed DNA sequences in this interval. Because this is a very tedious process, it is recommended that a very high-resolution, accurate, genetic map first be established. The closest flanking genetic markers that show recombination with the mutation are used to isolate YAC (yeast artificial chromosome) or BAC (bacterial artificial chromosome) clones that define the boundaries of the physical map. All markers that do not recombine with the mutation are also used to identify clones. In most cases, these clones do not overlap, and it is necessary to derive additional region-specific markers from them in order to isolate new clones, and repeat this cycle until a collection of overlapping clones, covering, with no gap, the entire region between the two limiting markers, has been obtained. Such a collection is known as a *contig*. All additional genetic markers produced during this process should be tested on the recombinant mice with the hope of further reducing the minimal genetic interval. Finally, coding sequences should be looked for in all clones from the contig, with the help of techniques such as exon trapping or cDNA selection. When a coding sequence differs between unaffected and affected mice, it is necessary to prove that it is responsible for the phenotype and is not a mere DNA polymorphism. This is easily achieved when this difference results in a premature stop codon or is a deletion and the sequence encodes a truncated polypeptide. When the difference results only in the change of one amino acid, the definitive proof that this is the mutation may come from three lines of evidence: (1) the existence of another mutation, allelic to the mutation of interest, with abnormal DNA sequence in the same gene, (2) the mutant protein has altered biological properties (reduced enzymatic activity, lower affinity for a ligand or receptor), or (3) rescuing the phenotype by transgenesis with wild-type DNA sequence (although negative results should be considered with caution).

In total, positional cloning of a mutation usually takes several years for a single person to identify the gene. The necessary skills are not only confirmed knowledge of molecular biology techniques but also expertise in biocomputing (including genomic databases and sequence analysis programs).

## VI. CONCLUSION

Genetic mapping of the causative mutated gene is an unavoidable step in the characterization of a new mutation. With the development of microsatellite markers, and because breeding schemes can be easily followed, it has become practical for almost any laboratory to do, even with limited experience in genetics. It can be predicted that within the next few years more and more mutations will be cloned through a candidate gene approach by taking advantage of both the cloning of hundreds of thousands of ESTs in the human and the chromosomal homologies identified between mice and men.

## ACKNOWLEDGMENTS

The author thanks Dr. J.P. Sundberg, Dr. M.T. Davisson, and D. Boggess for careful review of this manuscript. This work was supported by the Association Française contre les Myopathies and by EEC grant BMH4-CT97-2715.

## REFERENCES

1. Lyon, M.F., Epigenetic inheritance in mammals, *Trends Genet.*, 9, 123, 1993.
2. Lander, E.S. and Schork, N.J., Genetic dissection of complex traits, *Science*, 265, 2037, 1994.
3. Wilcox, A.S., Khan, A.S., Hopkins, J.A., and Sikela, J.M., Use of 3' untranslated sequences of human cDNAs for rapid chromosome assignment and conversion to STSs: implication for an expression map of the genome, *Nucl. Acids Res.*, 19, 1837, 1991.
4. Eppig, J.T. and Nadeau, J.H., Comparative maps: the mammalian jigsaw puzzle, *Curr. Op. Genet. Dev.*, 5, 709, 1995.
5. Rowe, L.B., Nadeau, J.H., Turner, R., Frankel, W.N., Letts, V.A., Eppig, J.T., Ko, M.S., Thurston, S.J., and Birkenmeier, E.H., Maps from two interspecific DNA panels available as a community genetic mapping resource, *Mammal. Genome*, 5, 253, 1994.
6. Flaherty, L. and Herron, B., The new kid on the block — a whole genome mouse radiation hybrid panel, *Mamm. Genome*, 9, 417, 1998.
7. Morse, H.C., *Origin of Inbred Mice*, Academic Press, New York, 1978.
8. Dietrich, W.F., Miller, J.C., Steen, R.G., Merchant, M., Damron, D., Nahf, R., Gross, A., Joyce, D.C., Wessel, M., Dredge, R.D., et al., A genetic map of the mouse with 4,006 simple sequence length polymorphisms, *Nat. Genet.*, 7, 220, 1994.
9. Bonhomme, F., Martin, S., and Thaler, L., Hybridation en laboratoire de *Mus musculus* L. et *Mus spretus* Lataste, *Experientia*, 34, 1140, 1978.
10. Bonhomme, F., Catalan, J., Britton-Davidian, J., Chapman, V.M., Moriwaki, K., Nevo, E., and Thaler, L., Biochemical diversity and evolution in the genus *Mus*, *Biochem. Genet.*, 22, 275, 1984.
11. Himmelbauer, H. and Silver, L.M., High-resolution comparative mapping of mouse Chromosome 17, *Genomics*, 17, 110, 1993.
12. Pavan, W.J., Mac, S., Cheng, M., and Tilghman, S.M., Quantitative trait loci that modify the severity of spotting in piebald mice, *Genome Research*, 5, 29, 1995.
13. Cox, R.D., Copeland, N.G., Jenkins, N.A., and Lehrach, H., Interspersed repetitive element polymerase chain reaction product mapping using a mouse interspecific backcross, *Genomics*, 10, 375, 1991.
14. Julier, C., de Gouyon, B., Georges, M., Guénet, J.-L., Nakamura, Y., Avner, P., and Lathrop, G.M., Minisatellite linkage maps in the mouse by cross-hybridization with human probes containing tandem repeats, *Proc. Natl. Acad. Sci. USA*, 87, 4585, 1990.
15. Weber, J.L. and May, P.E., Abundant class of human DNA polymorphisms which can be typed using the polymerase chain reaction, *Am. J. Hum. Genet.*, 44, 388, 1989.
16. Weber, J.L., Informativeness of human (dC-dA)n (dG-dT)n polymorphisms, *Genomics*, 7, 524, 1990.
17. Love, J.M., Knight, A.M., McAleer, M.A., and Todd, J.A., Towards construction of a high resolution map of the mouse genome using PCR-analysed microsatellites, *Nucl. Acids Res.*, 18, 4123, 1990.
18. Taylor, B.A., Navin, A., and Phillips, S.J., PCR-amplification of simple sequence repeat variants from pooled DNA samples for rapidly mapping new mutations of the mouse, *Genomics*, 21, 626, 1994.
19. Rhodes, M., Straw, R., Fernando, S., Evans, E., Lacey, T., Dearlove, A., Greystrong, J., Walker, J., Watson, P., Weston, P., et al., A high resolution microsatellite map of the mouse genome, *Genome Research*, 8, 531, 1998.
20. Avner, P., Amar, L., Dandolo, L., and Guénét, J.-L., Genetic analysis of the mouse using interspecific crosses, *Trends Genet.*, 4, 18, 1988.

21. Copeland, N.G., Jenkins, N.A., Gilbert, D.J., Eppig, J.T., Maltais, L.J., Miller, J.C., Dietrich, W.F., Weaver, A., Lincoln, S.E., Steen, R.G., et al., A genetic linkage map of the mouse: current applications and future prospects, *Science*, 262, 57, 1993.

22. Beier, D.R., Dushkin, H., and Sussman, D.J., Mapping genes in the mouse using single-strand conformation polymorphism analysis of recombinant inbred strains and interspecific crosses, *Proc. Natl. Acad. Sci. USA*, 89, 9102, 1992.

23. Manly, K.F., A Macintosh program for storage and analysis of experimental genetic mapping data, *Mammal. Genome*, 4, 303, 1993.

24. Lander, E.S., Green, P., Abrahamson, J., Barlow, A., Daly, M., Lincoln, S., and Newburg, L., MAPMAKER: an interactive computer package for constructing primary linkage maps of experimental and natural populations, *Genomics*, 1, 174, 1987.

25. Boguski, M.S. and Schuler, G.D., ESTablishing a human transcript map, *Nature Genet.*, 10, 369, 1995.

# 3 Computerized Colony Management

*David N. Larkins*

## CONTENTS

## I. INTRODUCTION

Most of the requirements detailed here for a computerized colony management system could be made applicable to managing colonies of any mammal, but as the focus of this book is around the research mouse, all references will be made to mice.

All development of colonies requires some form of management system, whether it is a few breeding pairs remembered by the mouse room technician, a paper ledger, a computer spreadsheet, or a custom-developed database system requiring a full-time maintenance staff. The paper ledger is an excellent system for ease of data entry, flexibility of data types, and addition of random comments. It also has significant disadvantages in that it is not easy to extract the data for electronic analysis, and the data are prone to interpretive error while in the documentation system. The spreadsheet is another effective system that attends to the analysis issues of the paper ledger, but it is prone to error on data entry, and the data organization in a single table becomes inefficient with larger data sets. The custom-developed database application is clearly the most attractive solution, with a specifically designed interface and relational database to exactly model the user's requirements. Unfortunately this system is out of the reach of most researchers for purely financial reasons.

If it were possible to develop a computerized management system with a feature set that would satisfy a wide range of users, the high cost of development could be amortized over the users, making the system much more affordable. Clearly, it is not possible to develop a system that would allow storage of every data item required by every potential user, so some sort of arbitration must be made in the design. Biological science has many common terms that are used to describe its data. For example, any mouse has a sex even if it is "other" and any mouse has a coat color even if it is "transparent." Other data types that share commonality are items such as phenotypes, genotypes, and procedures. These common data types make up much of the data set required in a colony management system. Using these common data types but including a flexibility of actual data would satisfy a large percentage of the user requirements. Adding a level of user configuration along with the flexible use of common data could make a system that would work for many different users.

A common software technique used for designing data-intensive systems is to first develop an abstract formation of the data referred to as a "model." As this model is refined, a physical representation of the data is developed to fit the model. A simple example of this is that regulations require no more than two breeding female mice to a cage, so the model defines a breeding cage as having up to two females in it. At a later time if a requirement is made for the system to be able to store three breeding females per cage, the model is referred to before any changes are made. In this situation if the new requirement does not fit the model, the new requirements should either be rejected or the model updated to reflect the changes.

There are also other more general considerations to developing a system that would be attractive to a large set of users, such as ease of operation, the use of existing hardware, and not requiring expert maintenance.

## II. THE MODEL

The basic requirement of the model is to describe an individual mouse in a colony of mice in a way that enables the user to store and extract as wide a range of pertinent data as possible. The model should be able to store all commonly required data concerning the mouse as well as sufficient user data to allow individual configurations to suit personal requirements. The model should allow filtering and sorting of the data and be able to present it in a fashion useful to either human or machine. The process used to establish the data requirements of this model was a series of interviews with scientific staff from several laboratories. This was chosen in favor of a questionnaire system to catch more detailed requirements.

### A. Basis

There are two common ways to look at mouse colonies, from a breeding unit or cage-centric viewpoint or from an individual mouse or mouse-centric viewpoint. Each has its own merits, as discussed in Strategies for Record Keeping,[1] and for effective mouse-specific information, the mouse-centric system was adopted.

### B. Database

The next step was to understand how individual researchers wanted to store their data. The common answer was "in a database," but actually meaning, "in some magical structure on my computer that can tell me everything I want to know about my colony without my entering any data." In reality, a database can take many forms, but it is usually some sort of file structure stored by a computer in some form of read–write memory. Some questions of concern were voiced regarding the location of these files: how secure would they be, where would they physically reside, and who else could access them. There was a strong requirement for absolute privacy expressed in many discussions, yet a clear need for global access was expressed by others. Clearly any database developed to store colony information for researchers must be portable, in that it could physically reside on a non-networked personal computer or reside on a central system machine with multi-user access. The key word here for the model is "portable."

## C. EXPERIMENT

Some researchers will breed a single colony that is to be the target of their research for years to come, while others will breed multiple separate colonies for specific experimental purposes. This indicated a requirement for the ability to support multiple colonies in a single database as needed. The mechanism chosen to represent this was the concept of "experiments." The basic unit of storage in the model would not be an individual mouse but a group of mice, and the name for this group of mice was to be an "experiment." Thus to store even a single mouse in the database, first an experiment had to be created and then the new mouse created in the experiment. This would allow researchers to create a single experiment and enter their entire breeding stock into it or to create multiple experiments, with each group of mice having their own purpose. The experiment level of the database also becomes a useful repository of other information regarding the entire experiment, such as the experiment protocol and user configurable items.

## D. MOUSE

Once an experiment has been defined, the operator must store individual mice and all their associated data in the database, each referred to as belonging to a particular experiment. When considering mouse data, imagine two different experiments, one where only mouse age and a single phenotype are required for 10,000 mice and one where 200 mice were to be genotyped at 1000 different markers. If the same database scheme (the arrangement of tables that store the data) is to be used for both experiments, would it be reasonable to allocate storage for 1000 markers in the second experiment for each of the 10,000 mice in the first? Obviously not. This would require the allocation of 10,000,000 unnecessary data locations. The data must be stored as efficiently as possible when the numbers become large, so this is where the need for a database with multiple related tables becomes apparent. Because of this requirement to render the database more efficient, only clearly mouse-specific data, such as pedigree number, coat color, sex, etc., are stored in the mouse section of the model. Others, such as genotypes, etc., are stored in their own tables. This partitioning of data also leads to another desirable effect, partitioning of functionality. As the model, and hence the application, became more complex, a way to simplify operation was needed. This was achieved by separating functionality into areas such as "mouse" where only mouse-specific data could be edited. In the final application, if the user only needs to enter mouse data, the other areas, such as "wetlab," can be completely ignored.

## E. MEASUREMENTS

As the model for the basic mouse was established, the model for all the experimental data associated with the mice fell into place. Each data item should be constructed as a related table or a table that is only used for a closely related set of data. This is referred to as the tables being "normalized," giving maximum storage efficiency to the database. This also gave another clear functional area, the "wetlab" for creating and editing specific experimental data.

## F. CAGES

As indicated previously, the design of this model was to be mouse-centric, concentrating on storing and accessing data relating to individual mice in the database. At the same time, the mice in the model were bred for the purposes of the experiment, and knowledge of cages, parents, offspring, etc., was required. The approach taken for the cage system was to consider three classifications of mice:

- Parents
- Unregistered mice are mice without pedigree numbers
- Registered mice are mice with pedigree numbers

A few simple rules were applied, such as parents must be registered mice, a mouse must be registered to take part in an experiment, etc. Then the cages themselves were considered, and four classifications determined:

- Breeding cages are cages with only parents resident
- Litter cages are cages with parents and offspring resident
- Holding cages are cages with both registered and unregistered weaned mice
- Experiment cages are cages with registered mice

With this set of data, although rules are loosely applied, many of the requirements of an efficient cage tracking system can be easily met.

## G. VIEW

Although inspecting the data does not affect the storage of the data model, the model must be easily accessible to inspection of any data stored in it. This means that the various tables that construct the database must be correctly referenced to each other to allow simple extraction of data from different tables at the same time. This is part of the database being considered "normalized."

The view aspect of the model is probably where the most advantage can be obtained from such a management system because of the filtering, sorting, and other selection processes that can be applied to the data.

## H. REPORTS

To complete the model requirements, there must be some facility for printing hard-copy reports for inclusion in papers, presentations, etc. The reports should be almost as flexible as the view and offer standard and customized versions as well as being able to sort and filter data in the report.

# III. THE DEVELOPMENT AND OPERATION ENVIRONMENTS

## A. PLATFORMS AND OPERATING SYSTEMS

There are three hardware platform families currently in common use around the biological community:

- PC
- Macintosh
- Sun/DEC/SGI, etc. (Mini/Workstations)

All three platforms support local graphical user interface (GUI) programming and are potential hosts of a colony management system. The platforms each run different operating systems:

- The PC commonly runs Windows 95/98 or Windows NT
- The Macintosh commonly runs MacOS
- The Sun/DEC/SGI platforms commonly run some variety of Unix

When deciding which platform/OS to select for the development environment and which platform/OS to target for the completed application, both the current usage of the different platforms and the trends of purchase of new platforms were considered. In the software development field, the PC has by far the largest presence, with a some 20:1 ratio between software developers coding for the PC and the Macintosh. This clearly indicates that more capable development software is available for the PC, and, as such, that should be the choice for the development environment. In the biological research field, the current ratio of PC to Macintosh is around 1:1, with more PCs in the commercial sector and more Macintoshes in the academic sector. It is also apparent that the rate of purchase of new PCs is much higher than that of new Macintoshes and that more of the Macintosh platforms are of the older, slower variety. There is little indication that the number of Unix platforms requiring such a system would warrant extensive development effort, but clearly there is a requirement for both PC and Macintosh support at the application level.

## B. THE DATABASE

The whole purpose of this application is to store data in a fashion that leaves it readily accessible to the user in a desired fashion. In the case of this example, by the end of the modeling phase this project contained 31 tables, with a total of 353 columns requiring storage. Each of these tables also requires some form of relationship address to be stored along with it. To perform these tasks in the application as defined would require a tremendous amount of additional software effort. An alternative to this approach is to use a commercial database, where a general table system and relationship pointers have already been developed and the transfer of data to and from the database is under the control of a clearly defined language.

The choice of a commercial database offers other advantages. The package has the guarantee of data integrity from a tried and tested system; other utilities, such as backup and replicate, come with the package; and commercial databases are generally able to accept multiple simultaneous users. The only significant disadvantages are the added complexity, the added load to the CPU, and the added cost, none of which outweigh the advantages.

## C. LANGUAGES

The language a program is written in is generally some human-readable form that can be converted by the "compiler" into codes that the target computer can translate into the application's functionality. The "source code" must contain instructions on how to handle every allowable control input and how to present every required output down to the finest line around a display window.

Languages have developed over many years from relatively simple forms, such as Fortran, PL/1, and Basic, through the more modern object-oriented versions, such as C++ and Object Pascal, to today's Rapid Application Development (RAD) suites, including Delphi, PowerBuilder, and Visual C++. In general, the lower the level of (and older) the language, the closer it is to the central processing unit (CPU) functionality. An example of this would be to draw a line on the screen with a low-level language. First, one would have to instruct the CPU to set a pixel on the screen to a particular color, then test if enough have been set and, if not, set the next pixel and repeat the test until the required line has been drawn. A similar function with a higher level language would be a single instruction: draw a red line 20 pixels long. The RAD languages have taken this concept a step farther in establishing a set of predefined controls that can be simply dropped onto a background and connected with relatively simple code to produce the required functionality. An example of this is to display a picture in a window, drop a picture control onto the window, size it correctly with the mouse, assign a bitmap file to the control, then run the application. The picture will be displayed! These RAD languages also support the concept of modern object-oriented design.

The commercial banking, finance, and accounting sector has provided much of the initiative to develop these RAD languages. A valuable side effect has been that the datawindow, editbox, listbox, button control, radio button, check box, and picture control are ideally suited to developing applications such as a colony management system. It is such a language, PowerBuilder, that has made developing a complicated colony management program possible. Even so, the development time needed to release such a program is in the order of 5000 software hours.

## D. CONNECTING TO THE DATABASE

As the decisions to use a commercial database and a RAD language have been made, consideration must be given to how the two pieces of software are to communicate with each other. First, consider the database software. If the user already has site licenses for a commercial database, it would be a benefit not to purchase another database, but integrate the preexisting system. Also, if there were any reason to change the database sometime in the future, it would be preferable if it could be done without massive changes to the application software. The database vendors have already addressed these requirements with the creation of a common database connection language. The standard query language (SQL) can be used to query any database supporting SQL for the described type of data, resulting in the ability for any compliant database to be integrated quite easily. To simplify things even further, an additional feature provided by the language or database vendors is a method of connecting to the different types of database either directly or over a network for SQL transactions. This is known as open database connectivity, or ODBC. These connections need almost no work by the application developer.

It should be noted here that the SQL connection between differing databases is not always totally transparent and certain modifications to the application may be required to support additional databases.

## E. NETWORKING

The need for a networkable system becomes apparent as soon as one looks at the diversity of workplaces in research. There is great interest in being able to enter data directly from the mouse room. Investigators are interested in viewing data from a server in their laboratory or from their office, and supervisors are interested in connecting to one of several of their employees' databases from a single location. At the same time, there is also a requirement for multiple users to access the same database, sometimes simultaneously. These capabilities are provided for largely by the ODBC connection to the database.

## IV. IMPLEMENTING THE MODEL

When the data model is completed, the development environment determined, and the language and database chosen, the next step is to begin the implementation of a workable graphical user interface.

## A. THE GRAPHICAL USER INTERFACE (GUI)

The main requirement of the GUI is that the entry of data into and the extraction of data from the database is as simple as possible, consistent with the flexibility of function as determined by the data model. Other less obvious but important factors are that the GUI must appear professional and relatively pleasing to the eye. It must be uncluttered, yet offer all the functionality required in a particular window. The need to jump from window to window to detail a single data set must be minimized. The overall operation should be as self-intuitive as possible.

## B. User Assistance

In areas such as the mouse room, where attention is focused more on handling the mice than on data entry, there is functionality that can be added to greatly assist in entering the data. The simplest addition is to provide a dropdown list box where the sex of a mouse is to be entered. So rather than typing the word "Female," a simple (computer) mouse click, drag, release can enter the string "Female" in the required window. Another feature that can assist in a similar fashion is a button alongside the weaning date that, when clicked, enters the current date plus some predetermined number of weaning days.

Along these same lines of user assistance, other facilities can be added, such as checking for all mice in a experiment that are due to be weaned on a particular date or during a specific week. Functions such as these can assist greatly in the reduction of human error in raising the colony.

## C. User Configuration

There are always unforeseen items of data that are required but have not been incorporated into the data model. To accommodate such requirements, extensive user note storage can be added throughout the application, and user configurable drop-down lists can be added for faster input.

## V. COLONY

All the previous modeling sets the guidelines for the development of the final application, but it does not create the design. From this point on, the requirements developed previously should be incorporated into a cohesive program design document. Then coding the application can be initiated. The following description details the development of the model into an actual application called "Colony."

## A. Putting it All Together

First, some of the controls used in the application need to be understood:

- An application's window is the area on the computer screen that displays the functionality of the application.
- A pop-up is a small window with a few controls or a question that needs to be addressed before the program operation can continue.
- A *save* button will usually be complemented by a *cancel* button or a pop-up window that allows the save to be cancelled.

- A tab control is something like the window version of a file folder, with tabs that, when clicked, can select one of the pages in the folder. One of the pages in a tab control is called a tab-page.
- A button is an area of the window whose graphics indicate a slightly raised surface that looks similar to a button on a physical control panel. It usually has a name indicating its function.
- An edit box is an area of the window whose graphics indicate a slightly depressed surface that can be used for numerical or text entry. If the box color is white, it is intended for data entry, if the color is the same gray as the window it is intended for display only.
- A listbox is an area of the window whose graphics indicate a frame around a display area that displays lists of data, such as a list of mice.
- A checkbox is an area of the window whose graphics indicate a small square. Clicking on this square will produce or erase a check mark in the square. The function of this control is to offer a yes–no decision.
- A radio button is an area of the window whose graphics indicate a small circle. Clicking on this circle will produce or erase a dot in the circle. The function of this control is to offer a one-of-many selection.

The starting point is to connect to the database so a simple window is constructed to enter a userID and a password, with OK and Cancel buttons to proceed or exit. Logging into the system presents an activity manager where the main functional choices can be made. As indicated previously, if no wetlab information is to be entered, there is no requirement to select the Wetlab button. As a little additional security here, switches were created in the setup file that can disable any or all of the activity buttons if required.

An experiment window is provided to contain information pertinent to the selected experiment. It is also able to store a protocol document for the experiment and any user-configurable drop-down lists data the user has added. On editing the experiment data, it can be saved to the database or cancelled, and the window closed.

A mouse window is provided to create new mice and edit mouse-specific data, such as coat color and sex, along with birthdates, etc. There is a second tab-page that is intended to provide information about the parentage of the mouse, its parental references, whether it was fostered, or whether it was hosted. Next, there is an ID tab-page intended to record information about the way the mouse is marked for ID, including toe-clip, ear-notch, tattoo, and implant data. There is a reference tab-page to point the database to any other database that may be used, such as histology or pathology databases. Finally, there is a view tab-page, allowing mice to be viewed en masse, without switching to the view window.

A cage window is provided to create and populate cages in the mouse room. It is important to note here that there is no magic connection between the database and the mouse room. If a mouse escapes, the computer will not know unless a human operator has told it. The cage model is based on three types of mouse: parents, unregistered pups, and registered pups. These three types of mice can inhabit up to four types of cage: the breeding cage, the litter cage, the holding cage, and the experiment cage. The cages tab-page simply creates or edits cages. The breeding tab-page allows the operator to assign parents to cages (maximum one male, two females), and the litter tab-page allows litters to be created in those cages. The litter to holding cage, holding cage, and experiment cage tab-pages allow movement of mice between the various cages. The movement of mice between different types of cages proved to be an excellent opportunity to use the drag and drop features common to windows programming. Tab-pages recording test matings and litter history are also included here.

A wetlab window is provided with the first and second tab-pages, defining, entering, and editing phenotypes and genotypes. Treatments and procedures can also be defined and entered and also contain an activity date. If an appropriate date for a treatment or procedure is entered, this treatment or procedure can be automatically reported as being due in a further part of the application. Because of the very specific issues of storing samples in freezers, a separate freezer tab-page is provided, and for other samples there is a general sample tab-page. One feature of this tab-page is that it is able to store images along with the rest of the data. This would enable a researcher to record the storage location of microscope slides for a particular mouse and also record the actual image from the slide in the database. This could be very useful for passing data between collaborators.

A view window is provided to view all data in the database. The first tab-page offers a general view of mice in the experiment. The next two tab-pages display all phenotypes and genotypes created. Then there is a cage view tab-page that allows a view of all cages and their inhabitants. This page will also allow the user to check for any mice due to be weaned this day, this week, or next week. The next tab-page offers a tree view of the colony ancestry or progeny. The data to generate these views are automatically generated if the breeding/litter cage system is used. There is a tab-page to check treatment or procedure activity, which will report any due dates this day, this week, or next week. Then there are two tab-pages to inspect the freezer and sample data present in the experiment.

Finally, there is a data window that can export data in a more specialized way than the normal ODBC connection and that is designed to prepare the data for input into other proprietary analysis tools.

## B. USING THE PROGRAM

Adding all the requirements of the model, the database, and the implementation together produces a fairly complex operational interface that may seem far more than is required for maintaining a colony tracking just sex and coat color. The application attempts to simplify operation by grouping functionality. To record just sex and coat color, once an experiment is defined, only the mouse window is required for simple creation of mice and selection of other parameters. Similarly, if only registered mice are to be entered in experiment cages, only the cages and the experiment cage tab-pages are required. If a complex experiment with mouse IDs, treatments, genotyping, phenotypes, and samples are to be recorded and the mice bred through the entire cage system, most of the application will be used.

# VI. FUTURE PLANS

## A. POSSIBLE NEW FEATURES

If the program is adopted as a viable colony management system, there are likely to be requirements for further features to be added to the existing set. Some potential additional features are discussed here.

### 1. Copy

A copy feature to allow selected mice to be copied to either a new experiment or to a new database is a very likely addition. This would provide support for pools, where a certain pool of mice would be selected from an experiment for further work. An example of this would be: after performing a genome wide scan, groups of mice exhibiting characteristics in areas of interest would be pooled into a separate experiment for further study. Copy would also provide support for a breeding room developing its own colonies and then copying selected mice to a separate database for distribution with the mice to the investigator.

### 2. Import

When an investigator is several years into an experiment that uses spreadsheets for colony management and he/she needs to move to a system such as this due to the increase of scope, discarding all previous data is not an option. Providing an import facility would allow smoother integration of the program into existing laboratories.

### 3. Filter

Due to time constraints, the current application does not provide filtering on multiple phenotypes or genotypes. This is an essential feature to provide in a further release for more advanced analysis.

### 4. Printing

A facility to print some of the specialized features that the normal report generator cannot would be a valuable addition. One example of this would be to provide a printed output of the ancestry/progeny tree window.

## B. POSSIBLE NEW MODULES

Differing from added features, an additional module would be a completely separate piece of software that could be loaded or not. Examples of such modules may be QTL and statistical analysis packages or graphical output systems.

## C. POSSIBLE NEW VERSIONS

Work on a smaller version just to handle mouse data has already been started to see if there is a requirement for a significantly simpler application. Another potential version may be focused more on the animal resource facility rather than the research laboratory.

## VII. CONCLUSION

The computerized colony management system offers the opportunity to increase the efficiency and reduce the error rate in even small mouse colonies. It also offers potentially far more powerful searching and matching capabilities. It does require discipline and diligence in application.

This brief description has been an attempt to outline the process of creating such a program and the results obtained from pursuing this process. The final application is now actively being used in several laboratories around the U.S. and being tested further afield.

Colony is copyrighted by The Jackson Laboratory and licensed for distribution to Locus Technology Inc. Details of its purchase can be found at: http://www.locustechnology.com.

## ACKNOWLEDGMENTS

I would like to thank the research staff at The Jackson Laboratory for helping me understand the mouse model.

## REFERENCES

1. Silver, L. M., *Mouse Genetics: Concepts and Applications,* Oxford University Press, New York, 1995, 57.

# 4 Medical Record Keeping for Project Analysis

*Beth A. Sundberg and John P. Sundberg*

## CONTENTS

## I. INTRODUCTION

Maintenance of records about breeding colonies and associated research is highly specialized and can easily become overwhelming if a systematic and proven method is not utilized (See Chapter 3, Computerized Colony Management). The same is true for medical records on each mouse necropsied, both to organize all data generated and to access various specimens collected and materials stored. Generation of data and case material is useless unless it can be retrieved and analyzed on demand. Computerized databases and spreadsheets provide invaluable tools for maintaining large amounts of data that can easily be processed to generate tables and graphics for summarization, interpretation, and presentation.[1]

The researcher must keep extensive records on the mouse colony, experimental test results, medical records, specimens, photographs, and other information. Many different methods are often used to maintain this information, including paper records on forms or in notebooks, computerized spreadsheets, computerized databases, automated laboratory equipment printouts, and specially designed computerized laboratory information systems. Additional information may then be created in the form of reports, statistics, graphs, posters, formal presentations, drafts of papers, and final publications. This information is frequently scattered in many locations and in the possession of the various people involved in the project. Sometimes this proliferation of information can create great difficulties in retrieving specific data if a key person is no longer present. Hence, it is highly recommended that the researcher establish a record keeping system at the beginning of the project that provides a consistent framework for use by all participants. Ancillary materials, such as glass microscope slides, $2 \times 2$ transparencies, paraffin blocks, etc., can be filed under a common identifier (case/accession number) to facilitate retrieval.

The project record keeping system will depend on the resources available to the individual researcher. Ideally, a single record keeping system should be used by everyone in the same research laboratory so that materials from different individuals' projects can be easily identified, sorted, and retrieved, even after they leave the group. Looking at the available facilities and establishing some protocols for information recording at the beginning of the project (or when establishing a research laboratory) will minimize problems later. Large automated (computerized) laboratory information systems may already be in place and need to be used. For many researchers, however, these systems are not generally available, may be available at a prohibitive cost, or cannot be adapted to a particular investigator's needs. This chapter will focus on the creation of your own specialized information system using available resources.

## II. SOFTWARE

Some phases of a project will require special software, while others may allow many choices. For example, image analysis may require use of a particular system and may limit the software this information can be output to. Proprietary format data may be included in the software package. Conversion of these data into a commonly used spreadsheet program or database may have to be done manually or special software might need to be acquired. On the other hand, the many available word processing software programs can provide too many choices. If all project participants are required to use the same word processing program, drafts can be exchanged with minimal loss of formatting.

Determining the bottlenecks or limitations of particular software in the beginning can avoid problems and delays later. The research team should establish the needs and flow pattern of the project, who will be gathering the data and putting it into computerized form, when and how test results will be collected, when analysis will begin, and what types of analysis will be needed. Do not forget the final presentation formats. Based on the goals for a final product, different software can be tested to answer those needs. Once the data flow pattern has been established, it can be used repeatedly for different projects and could even allow for the establishment of a historical database and specimen collection.

The basic types of programs needed are a relational database and a compatible spreadsheet into which selected data can be placed for analyses (Table 4.1). Statistical and other specialized programs for gene mapping that can utilize the spreadsheets generated provide additional tools for manipulation and analysis of data generated (Table 4.1).

## III. HARDWARE

Most information today eventually finds its way to some mixture of computers and computer software programs. Start by assessing how many different types of com-

**TABLE 4.1**
**Our Sources of Software for Data Maintenance and Analysis**

Word Processing Programs

| | | |
|---|---|---|
| Corel WordPerfect | Corel Corporation (U.S.A.) | (801) 765-4010 |
| | 567 East Timpanogos Parkway | Fax (801) 222-4379 |
| | Orem, Utah 84097-6209 | www.corel.com |
| MS Word | Microsoft Corporation | (425) 882-8080 |
| | One Microsoft Way | www.microsoft.com |
| | Redmond, WA 98052-6399 | |

Relational Database Programs

| | |
|---|---|
| FoxPro for Windows 2.6 | Microsoft Corporation |

Spreadsheet Programs

| | |
|---|---|
| MS Excel | Microsoft Corporation |

Presentation Programs

| | |
|---|---|
| MS PowerPoint | Microsoft Corporation |

Colony Management Programs

| | | |
|---|---|---|
| Colony | The Jackson Laboratory | www.locustechnology.com |
| | 600 Main St. | |
| | Bar Harbor, ME 04609 | |

Statistical Programs

| | | |
|---|---|---|
| StatView | SAS Institute Inc. | (415) 623-2032 |
| | Attn: StatView Sales and Marketing | Fax (415) 623-2083 |
| | Two Embarcadero Center, Suite 200 | www.statview.com |
| | San Francisco, CA 94111-3834 | |

Gene Mapping Programs

| | | |
|---|---|---|
| Map Manager XP | Robert Cudmore & Kenneth Manly | kmanly@mcbio.med.buffalo.edu |
| | Roswell Park Cancer Institute | mcbio.med.buffalo. |
| | Department of Cellular and Molecular | edu/mapmgr.html |
| |   Biology | |
| | State University of New York at Buffalo | |
| | Buffalo, New York 14263 | |

puters you are currently using. Are there compatibility problems between different brands (Macintosh vs. PC vs. others)? Is it possible for everyone to work on the same type of computer, or will the combination of software package needs require that information be shifted from one to another? Can this information be moved smoothly? For example, a Macintosh may be available in the mouse room, while a PC is used by the researchers in their offices or laboratories. Is it possible to use only one type of computer platform to alleviate compatibility problems?

Some software allows good transfer of information, so compatibility is not an issue. For example, an MS Excel spreadsheet file from a Macintosh computer can be transferred on a diskette or via an email attachment to a PC running an appropriate version of the MS Excel program.

If possible, software needs should be established first. Then fitting new hardware to those needs will allow you to select the best combination.

## IV. EXAMPLES OF A WORKING COMPUTERIZED SYSTEM

### A. STANDARDIZATION

### 1. Unique Identifier

Large scale data collection operations, such as diagnostic laboratories, clinics in hospitals, or drug safety studies in industry, assign a specific identifier for each individual animal. This is often referred to as a case or accession number and is utilized to organize all the materials generated by the case. In a relational database, this unique identifying number becomes what is called the primary key and is used to tie all the database information together.

A variety of simple numbering methods are used. The full year or part of the year number can be used as a prefix, or letters in the alphabet that correspond to a particular year can be used. For example, all cases that were handled during 1999 might be prefixed by 1999, 99, or A. This would allow all cases within a year to be searched or materials found in a pile in the laboratory to be quickly identified as to when they were generated and which file cabinet the materials should be stored in. Accessions are then numbered numerically, or groups of numbers can be assigned to particular projects, programs, etc. For example, we use five-digit codes in numerical order: 1 to 6999 are used for routine disease surveillance cases; 7000 to 8999 are used for diagnostic pathology cases; and numbers 9000 and higher are used for research-related cases.

Final assignment of the number can look like: A00001, 99-1, 9900001, 1999-00001, or some variation thereof. Using our system, this immediately tells you the year and the program that generated the materials in hand.

Assignment of numbers to animals is usually done in one of two manners. Since our laboratory runs multiple types of research and diagnostic programs, most of which focus on necropsies of mice from a variety of sources, we assign numbers to the mice when they come to necropsy. A lab that imports mice or uses only mice from within their own colonies may find that it is easier to assign the number to the animal when it is born or purchased. This is particularly true when using some of the computerized software for colony breeding management (Table 4.1, see Chapter 3, Computerized Colony Management). If a number is assigned when the mouse enters the colony/research laboratory, that number should be maintained for all data collected.

Complications in record keeping arise when specimens are shipped to collaborators or other services within the institution that have their own accession system. For example, specimens collected for ultrastructural studies are assigned a unique identifier and then sent to the Electron Microscopy (EM) Service for processing. The EM Service assigns its own accession number. When the investigator arrives to review the material with the electron microscopist, confusion is inevitable unless the numbers and data are cross-referenced. Solutions include adding the service's accession number to a field in the database for maintaining a cross reference or submitting the investigator's unique identifier and signalment for the case (animal type [mouse, rat], age, and sex), as well as information on the study to the service prior to examination. This is the type of data flow problem that should be noted and solved before the project begins. In our case, it is useful to arrive at the electron microscope with data sheets from our database in hand to assure no confusion will delay progress or confuse interpretation of observations.

## 2. Project Identifiers

A project name or short nickname is defined and entered for each mouse within that project. This project name allows searching through the database or spreadsheet to find all mice or other data collected for that particular project. If the project involves collaborators or another investigator, separate fields can be set up under submitter/collaborator to permit rapid searches for all work that involves this individual or group. Often multiple studies are ongoing with a particular investigator and summaries of all work are needed. This approach enables the program to set up the summaries quickly and, most importantly, reproducibly.

## 3. Other Standards

Other standardization suggestions include using the project name in addition to the accession number to label or categorize everything. Keep all information on a project in one folder in each computer. This folder can contain subfolders for different categories, such as spreadsheet data, word processing documents, final publications, statistics, etc.

## B. COLLECTION OF INFORMATION FROM THE COLONY

It may be possible to bring computers into the animal rooms to collect information directly. The use of bar codes and readers, balances tied into institutional networks to automatically record weights, etc., have become commonplace in mouse rooms. Specialized programs are becoming available for maintaining breeding records this way (see Chapter 3, Computerized Colony Management). If these data acquisition systems are not available, some type of paper form must be used for data collection. These data can then be entered later.

Our system normally collects individual mouse information into the database at the point the mouse is necropsied. An example of the data collection format is shown in Figure 4.1. The following signalment data are used:

*Case/accession number* (necropsy number): unique identifier
*Project identifier* (clinical number): to define the project
*Submitter/collaborator* (from): primary research contact for this mouse (PI, collaborator, other investigator, student)
*Location* (building and room): where the animal was housed (animal room number)

*Strain*: strain identification abbreviation

*Genotype*: usually mutant or control mouse information

*Pedigree number:* this may be assigned by the colony management program and serve as the accession number, or it may be a unique number assigned by the person maintaining the animals. Often pedigrees are used to track the previous generation.

*Other identifiers* (other ID): cage number, special coding, ear tags, etc.

*Sex*: F (female), M (male), FS (neutered [spayed] female), MC (castrated male)

*Birth date*

*Death/necropsy date*: used to calculate age

*Mating*: paired (male and female), trio (male and two females)

## C. COLLECTION OF NECROPSY INFORMATION

Necropsy information is added in two phases, both directly after initial necropsy and after the final diagnosis has been made. Additional data are added as generated. The initial history, clinical signs, and gross description are entered as a searchable text field of variable length. Later, comments about the final diagnosis are also entered as a searchable text field. Additionally, individual diagnoses are listed. These diagnoses may be searched by keyword. Using text fields for the diagnoses avoids the problems associated with trying to use and remember arcane codes.

## D. COLLECTION OF SPECIAL TESTS

Any tests that may have been run, such as microbiological tests on selected tissues, are also recorded for each mouse along with the results. Space has been allocated to maintain a short comment on each test if necessary. Standardized tables were set up with simple codes for the commonly used tests.

## E. IDENTIFICATION OF OTHER MATERIAL ASSOCIATED WITH EACH MOUSE (CROSS-REFERENCE INFORMATION)

Fields are set up to record in a yes/no fashion if the following material has been collected for each mouse: gross (whole animal) photographs (color transparencies), photomicrographs, black and white negatives (gross and/or photomicrographs), photo CD images, electron micrographs, histology slides, frozen tissue, radiographs, *in situ* hybridization, immunohistochemistry, gene mapping data, and many other types of tests and their results. These items are physically stored or filed by project and case number in order to be easily accessible.

## F. REPORT GENERATION

This database is used to output several standardized reports on the individual mice and projects as needed. The database is easily accessed by searching on various criteria. More important, typed final reports can be easily generated. If these reports are signed and dated by the investigator, they become legal documents. It is always a concern that medical records can be subpoenaed or that research material may

## THE JACKSON LABORATORY

Pathology Program, 600 Main St., Bar Harbor, ME 04609-1500

| John P. Sundberg, DVM, Ph.D., | James Miller | Richard S. Smith, MD, D.Med. Sci., |
|---|---|---|
| Diplomate ACVP | Senior Diener | Diplomate AAO |

From Sundberg                    Bldg. & Rm. AAA-999                    Date 01/01/98

Strain        C57BL/6J

Necropsy Number
Genotype    +/?                                                        1998-99999

Pedigree #   1234.567

Clinical Number
LAHsm-9999

Other ID    sample project

Date Born 06/10/97   # of animals 1   Sex F   Mating Pair

Items recorded include: Color photographs  B & W photographs  Photomicrographs
Electron micrographs  Radiographs  In Situ Prep.  Histology  Frozen Tissue  Photo CD

---

History and Clinical Signs -- Gross Description
Gross findings - sample.
Variable length description including symptoms and observations.

---

Special Tests       # Neg  # Pos  Serology  Organism
Listing of multiple tests that were performed  and the results with comments

---

Diagnosis -- Comments
Variable length text description of the diagnosis and any other information

---

Morphologic Diagnoses
Listing of multiple diagnoses

From Pathologist     Date Telephoned   / /   Preliminary Report   / /     Final Report   / /

---

**FIGURE 4.1** Example of a worksheet format commonly used in diagnostic laboratories. The database entry form can be set up in an identical manner to permit rapid entry. (From Sundberg, B.A. and Sundberg, J.P., A database system for small diagnostic pathology laboratories, *Lab Animal,* 19, 55, 1990. With permission.)

become important in patent processing. A signed, dated, and filed hard copy attached to the original hand-written worksheet provides an unalterable reference. This can become important in legal situations but also provides a record that can be checked for accuracy if a particular data point stands out as apparently being entered incorrectly. For example, when doing retrospective case series studies of frequency and epidemiology of particular types of tumors in inbred strains of mice, exceptionally young mice may show up in the tabulation of case information. Checking the original case sheet may indicate that the wrong year was entered for the birth or necropsy date. For mice with an average life expectancy in a research or breeding colony of about a year, rarely two or more, this alters the data significantly. Maintenance of the original record allows for verification of accuracy of data entry, assuming the original worksheet was carefully prepared or the information can be followed back from there to obtain the correct result.

## G. STUDY ANALYSIS

Many scientists and technicians are familiar with the use of spreadsheets as a laboratory analysis tool. Our database system is used to output a spreadsheet of selected data to be used for analysis. All database fields can be searched for and selected for output. This spreadsheet is then used for additional data collection of project specific data that do not need to be kept in the larger historical database. Analysis tools in the spreadsheet package include summarization, statistics, and graphs. Final copies of tables and graphs can be created that are of publication quality. Graphs and tables can be copied to word processing software for inclusion in papers or to presentation software for inclusion in 35 mm slides, overheads, posters, etc. Spreadsheets can be used as input to statistical programs and gene mapping programs for further analysis.

## H. LABELS

Computer programs can be used to generate labels for $2 \times 2$ transparencies, folders, images, negative sleeves, etc. Using the case accession number and project identifier on the labels helps with filing and permits rapid identification if more data need to be looked up.

## V. SPECIFIC SOFTWARE

There are a multitude of inexpensive software packages available that can be modified to investigators' needs following the guidelines provided above. These include database programs, spreadsheet programs, and statistics programs. Use a database for maintenance of raw data. A relational database will permit the greatest flexibility. Selected data, based on various criteria, can be output from the database for analysis to spreadsheets or statistical programs. The spreadsheet and statistics programs often include better summarization tools and graphics (charts for publication) than the databases. Many marker analysis programs today accept data from standard spreadsheet programs as input. Many different programs of these types exist, and it would be impossible to analyze all of them here. Table 4.1. lists the software programs we use for these functions.

## VI. CONCLUSIONS

Inability to maintain organized records limits the speed at which a study can be finalized and reports/manuscripts generated. Data collection based on a single accession number incorporated into a relational database enables centralized storage of all information on an individual, pooling of information in the form of summary reports, or copying to compatible programs. Spreadsheets, statistical packages, colony management, and gene mapping software that are compatible with the database make rapid analysis possible. Use of a single accession number for all materials collected from an individual provides ease of storage and retrieval of specimens. Retrospective studies can be done using both the stored data and finding the original material for verification, photography, etc.

## ACKNOWLEDGMENTS

This work was supported by grants from the National Cancer Institute (CA34196) and the National Institutes of Health (AR43801, RR08911).

## REFERENCES

1. Sundberg, B.A. and Sundberg, J.P., A database system for small diagnostic pathology laboratories. *Lab. Anim.*, 19, 55, 1990.

# 5 Necropsy Methods for Laboratory Mice: Biological Characterization of a New Mutation

*Melissa J. Relyea, James Miller, Dawnalyn Boggess, and John P. Sundberg*

## CONTENTS

## I. INTRODUCTION

"Necropsy" has roots in the Greek *nekros* (corpse) and *opsis* (view), literally meaning to look at the dead. This is the correct term for doing a postmortem examination on a nonhuman species. The word "autopsy" comes from the Greek *autopsia,* which literally means "seeing with one's own eyes"[1] and denotes viewing one's self. The term "autopsy" should be limited to performing a postmortem examination on humans. Regardless, this is a very technical procedure, and if done properly, it can yield a great deal of information. Proper handling of the animal, both antemortem and postmortem, is necessary to optimize results. Many molecular publications that

0-8493-1905-6/00/$0.00+$.50
© 2000 by CRC Press LLC

report mutations induced through targeted mutagenesis or transgenesis fail to adequately evaluate all organ systems, thereby missing numerous phenotypic deviations (lesions) that are critical to the evaluation of gene function. This chapter will review methods used for complete necropsies of laboratory mice.

## II. BIOLOGICAL CHARACTERIZATION OF A NEW MUTATION

It is often thought that, since individual mice of inbred strains are essentially the same and that normal gross anatomy and histology would be the same for all mice, only a few mutants from a new strain need to be examined and that controls are not necessary. Similarly, there are those who believe that only the obviously affected organs or specific organs, such as the liver, kidney, and lungs of the mutation, need to be examined and that the whole mouse need not be looked at. Characterization of a new mutation using these limited approaches can lead to incomplete and erroneous descriptions of the phenotype. The only way to provide complete characterization of the phenotypic effects of a new mutation is to do a thorough, systematic evaluation of sufficient mutant and control mice. This can even be done when only a small breeding colony of the animals can be maintained, if mice are collected at specific ages, and specimens are accumulated over time.

There are many major changes in the life span of a mouse, as in other species, which can have an affect on the phenotype caused by a mutation. These life changes provide focal points for setting up a systematic evaluation. The major life periods we look at when characterizing a new mutation are birth, weaning (3 to 4 weeks), sexual maturity (6 to 8 weeks), sexual quiescence (6-8 months), and geriatric stages (1 to 2 years +). These age ranges allow for differences between strains.[2] In most instances the cost for maintaining mice through geriatric ages is prohibitive and since many mice with genetic mutations often do not live that long, this age group is rarely studied.

During the first three weeks of a mouse's life the skin and hair undergo some very dramatic changes. The epidermis of a normal newborn mouse starts out relatively thick and becomes thinner by two weeks of age.[3,4] The hair follicles develop completely during the first week after birth and the first hair fibers begin to emerge at around five days of age.[4] The hair cycle is synchronized and short (about 1 week intervals) during the first three weeks after birth, which makes it easy to study. It is important to collect skin from mice at two- to three-day intervals (birth, 3 days, 6 days, 9 days, 12 days, 15 day, 18 days, 21 days) during this time so that you may evaluate all of these features. Other organs undergo similar developmental changes during this period.[5] Interval-specific postpartum collection ages can be defined for each organ following this approach.

It is easy to imagine that the numbers of mice in this type of study could get quite large. However, since most mutations occur or are induced on inbred mouse strains maintained in controlled environments free of specific pathogens, it is possible to do a complete study on as few as two mutants and two controls of each sex at each of the time points mentioned above (8 total per age group). The mice should be housed in a clean, controlled, pathogen-free facility to ensure the accuracy of the study. Mice assembled for the major life change points should have complete sets of organs collected for study. This includes selected skin, spinal column, skeletal muscle, brown and white fat, femur, stifle joint, feet (to evaluate nails as well as bones), mammary gland, brain, liver, spleen, kidneys, pancreas, intestines, stomach, adrenal, lung, urinary bladder, trachea, thyroid gland, esophagus, genital tract, heart, tongue, lymph nodes, salivary gland, and preputial or clitoral glands. Mice collected for skin or other specific organ evaluation need only have the selected organ collected. For skin, this includes samples from the dorsal and ventral trunk, eyelids, ear, muzzle, tail, and footpads. Injecting the mice with bromodeoxyuridine or tritiated thymidine prior to euthanasia makes it possible to use multiple biopsies from individual mice for kinetic analysis, further reducing the numbers of animals needed to complete the study. However, mice with some skin mutations exhibit a positive Koebner reaction following injury, so the quality of multiple biopsies over time from the same mouse may be less than optimal.[6,7]

## III. CLINICAL EVALUATION

Live mice should be carefully examined for both behavioral and physical abnormalities. Most homozygous recessive mutations ($m/m$) are available with heterozygotes ($+/m$) or wild-type ($+/+$) age- and sex-matched controls on the same genetic background (where "$m$" = the mutant gene being studied, and "$+$" = the normal, or wild-type, gene).   Controls should be examined side by side with mutants as a basis for comparison. Familiarity with the normal phenotype of the background strain is essential in assessing the phenotypic variances of the mutant, and is also important in determining the presence of disease. Many infectious diseases in mice can present as behavioral abnormalities, such as circling or torticollis associated with middle ear infections. If these are evident in both mutant and control mice, it will be important to do infectious disease surveillance studies on the colony. Some mutations have clinical phenotypes that resemble infectious diseases, such as cutaneous scaling (possibly ringworm) or focal loss of hair (alopecia potentially due to ectoparasite infestation). Clinical and diagnostic veterinarians are able to deal with these effectively. Disease issues are beyond the scope of this book and are covered in detail elsewhere.[5,8]

Mice should be allowed to walk around in their box so that their patterns of behavior can be observed. The mice should constantly explore their environment and be alert and active. They should respond to external stimuli without abnormal reactions. Some inbred strains, like DBA/2J respond to loud noises by developing sudden and sometimes prolonged seizures often ending in death.[9] The mouse should have a uniform hair coat that lays flat. Vibrissae, the long hairs around the eyes and muzzle, should be straight and prominent. Ears should be erect and light pink in color for albino mice. Pale white ears in albino mice may suggest the presence of anemia. Eyes should be bilaterally symmetrical and clear. Nails should be short and curved. Incisors are commonly overgrown in many strains (malocclusion of incisors), but this may be part of the phenotype if it is constantly observed in association with a particular mouse mutation. Body openings should be checked to make sure that they are normal and normal secretions or excretions are produced. For example, normal mouse feces are about the size and shape of a rice grain, firm to hard in consistency, and dark brown in color. Perianal matting of feces or light yellow colored feces might indicate the phenotype of inflammatory bowel disease or the presence of any number of intestinal infections.[10] Several commonly used inbred strains of mice may have perianal swelling, giving the appearance of extra testicles (3 or 4) due to cysts of the bulbourethral glands.[11]

The animal identification information should be collected at this point. This includes age (birth date and necropsy day on records), sex, strain, genotype if known, pedigree numbers, animal identification numbers including ear tag or punch (Figure 5.1), or toenail amputation, source (room), body weight, reason for submission, and name of submitting technician or scientist. If the mouse is part of an ongoing study, a special code number for that study should be assigned to keep the records straight. (See Chapter 4, Medical Record Keeping for Project Analysis).

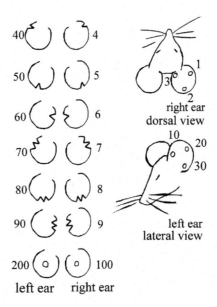

**FIGURE 5.1** Mouse identification by ear notching.

These are starting points. Thorough evaluation of the first mutants in a study will provide guidelines on what to look for more specifically as the study progresses.

## IV. CLINICAL PATHOLOGY

Routine collection of biological materials is done as part of physical examinations for humans and domestic animals in sickness or health. The methods are identical for mice but microassays are used. These can be done prospectively, as a routine procedure throughout all studies, or retrospectively, once a series of abnormalities is identified, to follow or define the pathogenesis of the disease. Since mutant mice usually provide a readily renewable population to study, the latter approach is commonly used in most research laboratories.

## A. BLOOD COLLECTION

Blood is collected by retro-orbital bleeding, tail tip amputation, cardiac puncture, or decapitation. The reasons to use each method vary with age, purpose of the study, volume needed, etc. Methods are summarized below.

### 1. Retro-Orbital Bleeding

This method is a valuable, nonterminal method of blood collection, normally used for small volumes of blood. The tip of a hematocrit tube is inserted into the medial canthus of the eye. Using a twisting motion with slight pressure on the tube, the retro-orbital venous plexus is ruptured. Once the blood is released, the hematocrit tube should be withdrawn slightly and the blood collected by capillary action. Retro-orbital bleeding is usually performed with manual restraint. Mice are grasped with thumb and index finger behind their ears, with their tail secured between the handler's last finger and palm to immobilize them. Some institutions require a variety of anesthetics or analgesics when performing this procedure.

### 2. Tail Bleeding

A tail bleed is accomplished by severing the lateral caudal vein on the ventral side of the base of the tail with a razor blade. This is often done in a specially constructed box into which the mouse is placed with its tail extending through an opening with an Eppendorf or glass tube for blood collection positioned below it (Figure 5.2). Some technicians may be able to hold the mouse as described for the eye bleed, for a manual tail bleed procedure. Care must be taken not to allow contamination of blood with fecal matter or urine when using the tail bleed method.

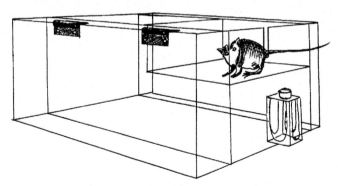

FIGURE 5.2 Plexiglas apparatus for tail bleeding mice. Mouse is placed in small, upper chamber with its tail extending through a slot. The ventral aspect of the base of the tail is cut with a single-edged razor blade. Blood is collected in a tube immediately below. (Drawn by Ingrid K. Sundberg.)

### 3. Cardiac Puncture

Cardiac puncture is a clean and thorough way to get as much blood as possible from a mouse at the time of necropsy. It must be performed immediately after the mouse is euthanized. Many people perform a cardiac puncture without opening up the thorax, by inserting a 20- to 22-gauge needle through the skin, intercostal muscles, and heart, then drawing blood into a syringe (a 1-cc syringe is usually large enough for most mice). We have found that carefully opening the thoracic cavity to reveal the heart results in cleaner needle insertion, allowing a larger volume of blood to be collected. This is achieved by following normal necropsy procedures outlined below, being careful not to sever any major blood vessels. Once the abdominal muscles have been incised, use a pair of sharp/sharp scissors to cut the ribs in a triangular shape, starting equidistant from the xiphoid process on both sides, and angling up toward the clavicle, again being careful not to sever any major vessels. Do not cut all the way through the sternum, as this would sever the thoracic aorta, causing blood loss. Fold back the rib cage to reveal the heart and lungs. With this method, the right ventricle may be cleanly punctured and the blood drawn slowly into the syringe. Gentle pressure applied to the internal organs, working toward the heart, pushes visceral blood into the ventricle for optimal collection. Blood should be promptly expressed into an Eppendorf tube before clotting occurs.

### 4. Decapitation

This method is seldom approved by Institutional Animal Care and Use Committees (IACUC) unless there is good reason. It is the method of choice for some endocrinological assays when hormones may be released during handling, thereby altering results to artificial levels. A number of commercial guillotine-type instruments are available (Harvard Apparatus Inc., Holliston, MA). Alternatively, mice can have their heads disarticulated with a large pair of scissors, (6.5 to 7 inches, blunt/sharp). Grasp the mouse firmly by the skin at the back of its neck and, holding it over a blood collection tube, swiftly and cleanly sever the head.

### B. Blood Handling

For serum collection, blood can be held at room temperature for an hour and then centrifuged. The serum should be decanted and stored in plastic tubes, (Eppendorf, Brinkmann Instruments, Inc., Westbury, NY; Nunc, Nalge Nunc International, Fisher Scientific, Pittsburg, PA) and frozen at -80°C until used. Blood may sit for several hours before centrifuging, if stored at 4°C.

Plasma is obtained by collecting blood in tubes containing EDTA or heparin to prevent clotting. The blood is then centrifuged and frozen for future use.

### C. Feces

Feces are collected at the time of necropsy and can be frozen in plastic tubes for a variety of assays. Most mice defecate upon handling so a few fresh samples are provided from a defined individual. This is a simple resource for *Helicobacter* spp. surveillance, using polymerase chain reaction methods.[12]

## D. URINE

Urine is usually expelled when a mouse is picked up. A handler can pick up the mouse as described in the retro-orbital bleeding procedure, holding the tail back with the little finger. In this position most mice micturate a few drops of urine. The urine can be collected in a plastic tube or tests may be done directly with a variety of tape strips. Chemstrip (Boehringer Mannheim Diagnostics, Indianapolis, IN) and Ames Multistix (Miles Inc. Diagnostic Division, Tarrytown, NY) are two urine analysis reagent strips that test for numerous urine components including glucose, ketones, protein content, and pH. For more specific tests that require larger volumes of urine or urine collected over defined intervals, metabolism cages are commercially available (Harvard Apparatus). Specific gravity of urine may be taken with a hand-held refractometer. Some companies offer refractometers made especially for urine testing, such as the Fisherbrand UriSystem Refractometer (Fisher, Pittsburgh, PA).

## V. GROSS PATHOLOGICAL EXAMINATION

Abnormalities are phenotypic deviations from known genetically based traits. Any abnormalities should be described on the records in as much detail as possible. Anatomic and pathological terms should be used, but careful descriptions using lay terms can be adequately translated by a medically trained collaborator. For example, "a tumor on the leg" provides little information. However, the statement that "a subcutaneous, hard, bony tumor, 1 cm in diameter, is located on the left rear leg near the pelvis," provides plenty of information that can be interpreted by a pathologist.

Simple and specific descriptive terms are commonly used by pathologists and are summarized in Tables 5.1 and 5.2. Combined with detailed anatomical localization, these descriptions provide a great deal of information on lesions in the mice. Mouse gross anatomy is detailed in several books.[13,14]

---

**Table 5.1**
**Basic Information Needed for All Necropsy Worksheets**

Signalment (species/mice, breed/strain/pedigree, color,
   sex, age[birth date], weight, animal identification number
Clinical history
Laboratory data (clinical chemistry, special tests)
Time of death (if submitted dead)
Mode of death (method of euthanasia)

---

## VI. FIXATIVES

A variety of chemical solutions are available that can preserve tissues for histologic, immunohistochemical, and ultrastructural studies (Table 5.3). Each has advantages and disadvantages. The choice depends upon the goals of the study, the tissue processing available in the histology laboratory your group uses, experience, and

---

**Table 5.2**
**Basic Components of a Description**

Organ or tissue name

Specific site (i.e., duodenum versus small intestine)

    Medial or lateral

    Dorsal or ventral

    Cranial or caudal

    Site specified by anatomic proximity (i.e., lumbar versus thoracic spinal cord)

Pattern and/or number

    Focal—circumscribed process

    Patchy—alterations that are multiple and poorly delineated

    Multifocal—indication of the specific number of foci contributes to the visual
      picture

    Diffuse - total involvement of the structure

Specific alteration and/or morphologic diagnosis (i.e., hemorrhages, abscesses,
  edema, pneumonia)

    Color

    Shape

Size and /or severity

    Enlargement or decrement - if uniform size change

    Degree - mild, moderate, severe

Etiology

    Gross examination - identify parasites if present

    Impression smears - identify infectious agents by appearance

---

the preference of the investigators. The solutions should be prepared in advance and be available in adequate volume and containers before the procedure begins. There are two general rules for histology: (1) Fixatives have various penetrating abilities and should be individually tested before use. Penetration is usually from 1 to 2 mm on any cut surface, so specimens should generally be trimmed and kept small; (2) For optimal preservation, use approximately 20 times the volume of fixative to the volume of tissues. Excessive amounts of blood and feces will limit the usefulness of the fixative. At the end of the necropsy, the fixative can be removed and replaced with fresh fixative to minimize or eliminate this problem.

Examples of commonly used fixatives for mouse histopathology are described below.

## A. Fekete's Acid Alcohol Formalin (Tellyesniczky/Fekete)

This is a commonly used fixative in mouse laboratories.[15-17] It provides rapid and surprisingly deep tissue fixation. Specimens are transferred to 70% ethanol after overnight fixation after which they are trimmed and processed. Long-term storage can be a problem because ethanol is flammable and evaporates easily. Once specimens dry out, they are useless. This fixative yields high-quality specimens in histologic sections. A disturbing artifact is that erythrocytes (red blood cells) are leached so that they appear only as pink ghosts within vessels.

## Table 5.3
## Formulae for Commonly Used Fixatives

**Tellyesniczky/Fekete:**

| | |
|---|---|
| 70% ETOH | 100 ml |
| Glacial Acetic Acid | 5 ml |
| 37-40% Formalin | 10 ml |

**Bouin's Solution:**

| | |
|---|---|
| Sat. Aq. Picric Acid | 85 ml |
| Glacial acetic acid | 5 ml |
| 37–40% Formalin | 10 ml |

**10% Neutral Buffered Formalin:**

| | |
|---|---|
| 37–40% Formalin | 100 ml |
| Distilled Water | 900 ml |
| Sodium Phosphate—Monobasic | 4 gm |
| Sodium Phosphate—Dibasic | 6.5 gm |

**B5 Fixative:**

| | |
|---|---|
| Mercuric Chloride | 6 gm |
| Sodium Acetate-Anhydrous | 1.25 gm |
| Distilled Water (hot) | 90 ml |
| Just before use, add: | |
| 37–40% Formalin | 10 ml |

**Carnoy's Fixative:**

| | |
|---|---|
| Absolute ETOH | 60 ml |
| Chloroform | 30 ml |
| Glacial Acetic Acid | 10 ml |

**4% Paraformaldehyde Fixative:**

| | |
|---|---|
| 16% Paraformaldehyde | 10 ml |
| Phosphate Buffered Saline (PBS) pH 7.2 | 30 ml |

**Glutaraldehyde Fixative:**

| | |
|---|---|
| 25% Glutaraldehyde | 10 ml |
| 0.2 $M$ Phosphate Buffer pH 7.2 | 50 ml |
| Distilled Water | 40 ml |

**Karnovsky's Fixative:**

| | |
|---|---|
| 25% Glutaraldehyde | 8 ml |
| 16% Paraformaldehyde | 12.5 ml |
| 0.2 $M$ Phosphate Buffer pH 7.4 | 50 ml |
| Distilled Water | 29.5 ml |

**JB4 Fixative:**

| | |
|---|---|
| 25% Glutaraldehyde | 16 ml |
| 16% Paraformaldehyde | 12.5 ml |
| 1 $M$ Cacodylate Buffer | 20 ml |
| Distilled Water | 151.5 ml |

## B. Bouin's Solution

This fixative uses picric acid (stains everything yellow permanently), acetic acid, and formalin.[16,17] Delicate detail is well preserved, but although commonly used for mouse tissues, it does not yield high-resolution specimens of fibrillar structures such as collagen. These structures present as hyalinized areas of amorphous pink material. Penetration is moderate. Bouin's fixed specimens must be washed in running tap water for 2 to 4 hours after initial overnight fixation and stored in 70% ethanol or they become very brittle. Bouin's solution can be used for decalcification of bones, since the acids will demineralize specimens that are left in the fixative for several days or weeks. If used for decalcification, the Bouin's solution must be changed once a week in order to optimize demineralization.

## C. Neutral Buffered 10% Formalin

This is the most commonly used fixative in many pathology laboratories.[16] Specimens can be fixed overnight and left in the fixative indefinitely. Specimens can be processed at any time as long as they remain wet, sometimes many years after initial collection. This fixative is particularly useful for retrospective evaluation of lipids or other substances that are soluble in ethanol and would be lost during tissue processing. For example, the presence of fat in adipocytes can be demonstrated if you take wet tissue fixed in neutral buffered 10% formalin, trim the tissue, cut frozen sections, and stain the sections with oil red O or some other lipid histochemical stain. Most of the other fixatives are alcohol based and remove lipid immediately.

Long-term storage in neutral buffered 10% formalin results in continued cross-linking of amino groups, resulting in changes in the tertiary structure of proteins. As a result, many epitopes are lost, making immunohistochemistry problematic. Transfer, after 12 hours, into 70% ethanol can reduce the problem.

Neutral buffered 10% formalin is noxious and should be used in a fume hood. Buffering is needed to reduce acid hematin formation, an artifact of fixation.

## D. B5 Fixative

This fixative is used for immunohistochemistry and often yields the most accurate and reproducible results of any fixative. It is difficult to use because it has to be prepared immediately before use, is based on mercury salts (difficult to dispose of), and fine precipitates can form (particular problems when using Gomori's methenamine silver or similar stains). Tissues fixed in B5 may be difficult to cut, and sections require treatment with Lugol's Iodine to remove pigments.

## E. Carnoy's Fixative

Rapid fixation of tissues occurs with this fixative. It preserves glycogen and enhances the staining of mast cell granules. Nissle granules are also well preserved. Because it is an alcohol/acid based fixative, it lyses red blood cells and acid soluble granules.

## F.  4% PARAFORMALDEHYDE

This is used as a fixative for electron microscopy and *in situ* hybridization. It should be mixed in a buffered solution at pH 7 and refrigerated. It will keep for several weeks this way.

## G.  GLUTARALDEHYDE

This fixative is commonly used for ultrastructural studies. Tissue penetration is very minimal, approximately 1 mm on any cut surface, so specimens have to be finely minced with a sharp razor blade to achieve adequate fixation. Several different buffers can be used. The most common are phosphate and cacodylate-based buffers. Phosphate buffers are safe and yield good results if used fresh. Very fine, electron-dense precipitates form that will render a specimen useless if the buffer used is old. Cacodylate buffers are arsenic based, which is toxic and can be difficult to dispose of properly. These are discussed in greater detail in Chapter 9, Ultrastructural Evaluation of Mouse Mutants.

## H.  KARNOVSKY'S FIXATIVE

This fixative is used for plastic embedding and electron microscopy. Karnovsky's is a general term for any fixative combining glutaraldehyde and paraformaldehyde in a phosphate buffer. Glutaraldehyde has minimal penetration ability. Paraformaldehyde penetrates deeper, but fixation is unstable. Karnovsky's fixative combines the positive points of both these chemicals.

## I.  JB4 FIXATIVE

This fixative is used for plastic embedding. It combines glutaraldehyde and paraformaldehyde in a cacodylate buffer.

## J.  O.C.T. COMPOUND

O.C.T. Compound (Tissue-Tek, Sakura Finetek U.S.A., Inc., Torrance, CA) is an embedding medium for frozen sections. It is a thick, clear fluid used in conjunction with plastic base molds such as CMS Tissue Path Disposable Plastic Base Molds (Curtis Matheson Scientific, Inc. Houston, TX) to bind fresh tissues for freezing and sectioning.

## VII. EUTHANASIA

Laboratory mice are usually provided live for necropsy. A variety of euthanasia methods are available and approved by the American Veterinary Medical Association.[18] Care should be taken to ensure that humane treatment is provided. Commonly used methods include:

## A. CARBON DIOXIDE ASPHYXIATION

This is a rapid and humane form of euthanasia for mice over the age of seven days. It is accomplished by utilizing a container designed to allow gas to enter rapidly and remove room air (Figure 5.3). These can be easily manufactured out of Plexiglas sheets and tubing or a large restaurant-sized clear glass jar may be used. Adequate ventilation should be available for the dieners performing the procedure. Gas is provided from a compressed gas cylinder attached to a wall or cabinet. The container is lined by a disposable plastic bag and preloaded with carbon dioxide gas by opening the valve on the attached cylinder to fill the container. The mouse is placed on the bottom of the container and the unit is refilled with carbon dioxide gas. The mouse will die within 1 to 2 minutes. Only one mouse should be euthanized at a time. Multiple mice placed in the container at the same time may result in those at the bottom not being killed. When the bag is removed and disposed of with the mice still in it, the underlying mice may revive. This is clearly an inhumane situation and should not be done.

Neonatal mice can be placed in these containers and will appear to be killed during the same time interval as for adults. However, neonatal mice are not euthanized by the gas unless it is done as follows: Place the pups in a plastic bag and

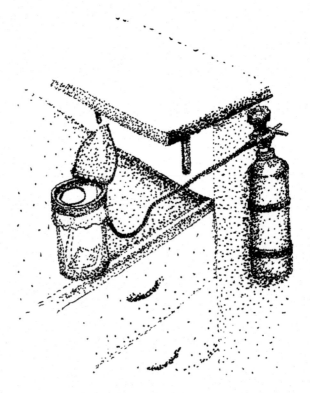

**FIGURE 5.3** Plexiglas container attached to a $CO_2$ cylinder for euthanizing mice. (Drawn by Ingrid K. Sundberg.)

without squeezing them, close down the bag around the pups and the inlet gas tube of the $CO_2$. Allow the bag to refill with $CO_2$ while maintaining a loose grip on the neck of the bag against the tube. Then close down the bag against the pups and the tube and refill once more. Repeat this for a third time. This assures the atmosphere in the bag is virtually 100% $CO_2$. The rate of gas entry into the bag should be slow enough so the gas temperature is not so cool that it chills the pups and makes them quiescent, thereby fooling you into thinking they have died. The pups soon become motionless and lose their color. Pups should routinely be decapitated with a pair of sharp scissors immediately upon removal from a $CO_2$ filled container to ensure euthanasia.

Dry ice is the solid form of carbon dioxide. It undergoes sublimation to carbon dioxide gas. Dry ice should not be used to euthanize mice. Adequate amounts of gas cannot be generated to do the procedure rapidly and humanely. Furthermore, if chunks of dry ice are put in the bottom of the container the mice can be injured by the low temperature of the material.

## B. BARBITURATE OVERDOSE

This is an effective and humane method that is described in detail under perfusion methods below.

## C. DECAPITATION OF ADULTS

This method should be avoided unless the experiment has very special requirements and special approvals have been obtained from the IACUC. For a description of this procedure, see section on blood collection.

## D. CERVICAL DISLOCATION

This method involves separating the vertebrae in the cervical area with a firm pinch to the neck and a pull of the tail. It is a quick, efficient method often used for routine diagnostic work. However, this method is not recommended for research mice, since it results in damage to tissues in the cervical area, as well as releasing large amounts of blood into the field, which can make observation and collection of some organs more difficult.

## VIII. PERFUSION METHODS

Perfusion combines euthanasia with fixation, providing the quickest way to get organs into fixative, resulting in the freshest tissues for study, with no degenerative changes. Special protocols must be followed to ensure this is done in a humane manner.

Materials needed include two 10-cc syringes with 23-gauge needles, one 1 cc syringe with a 23 gauge needle, stock pentobarbital sodium solution (50 mg/ml) (Nembutal, Abbott Laboratories, North Chicago, IL), phosphate buffered saline (PBS), fixative of choice, and 0.85% saline.

First, prepare the working solution of Nembutal by diluting 1.6 ml of the stock solution with 8.4 ml PBS. Then fill one 10 cc syringe with 0.85% saline, and the other with your chosen fixative. Label the syringes clearly to avoid confusion during the procedure.

The mouse must be anesthetized using an intraperitoneal injection of Nembutal working solution (0.1 ml/10 g body weight) using the 1-cc syringe. After the mouse is completely anesthetized, dip it in a mixture of water and disinfectant and pin it to a dissection board, ventral side up (See Section IX, Necropsy Procedure). Trim back the ventral skin from the thorax to the mandible to reveal the jugular veins that are located under both salivary glands. Following the necropsy procedure, immediately open the thoracic cavity. If a blood sample is needed, a heart puncture must be done at this time (see section on Blood Collection). After blood is collected, carefully cut the jugular veins. Blood will start to flow from the vessels. Insert the needle of the syringe containing the saline into the left ventricle of the heart. With steady pressure, perfuse the saline through and observe the area where the blood vessels were severed. The saline will wash the blood out of the system. It is important that the pressure exerted on the syringe is enough to push the blood out of the system, but not so much that it causes damage to any of the organs. After injecting 4 to 8 cc of saline, depending on the size of the mouse, repeat the same procedure with the fixative. If fixation is successful, the body will stiffen from the tip of the tail to the nose and all organs will blanch.   Tissue collection may then proceed as usual.

## IX. NECROPSY PROCEDURE

Once the mouse has been euthanized it should be superficially disinfected by submersion in a dilute solution of a germicidal detergent such as Calgon Vestal Process NPD One Step Germicidal Detergent (ConvaTec, St. Louis, MO), or a solution of 95% ethanol. When necropsying a mouse with an abnormality of the hair coat, it is important to collect samples of the hair before the mouse is dipped in the disinfectant. The hair should be plucked manually using the thumb and forefinger. Do not use forceps, as this will damage the hair shaft. Plucking of the hair allows for examination of the whole hair shaft from root to tip. The hair should be stored in a clean Nunc cryopreservation tube (Nalge Nunc International, Denmark). When studying mouse mutants, standardization of collection techniques is as important with hair as with other organs. Hair samples should be collected from the same area on every mouse in a study. We pluck the hair from the left flank from shoulder to hip to get hairs from both the dorsal and ventral surfaces. Be sure to take any skin that will be collected from an area of the mouse that was not plucked to prevent artifactual changes in the hair follicle being examined. If the vibrissae are abnormal, these should be collected as well, from the same side as the hair is plucked. They should be stored in a separate container.

If a mouse has a skin abnormality, remove hair prior to disinfection. This is usually accomplished by shaving the mouse with electric hair clippers such as the Oster Finisher Trimmer (Cat. # 76059-030, Oster Professional Products, McMinnville, TN). These clippers are easy to handle and have a small blade that is ideal for mice or other small mammals. If complete hair removal is desired, there are commercially available depilatory products (Nair, Carter-Wallace Inc., New York, NY; Neet, Reckitt & Coleman Inc., Wayne, NJ) that can be applied to the mouse after shaving. These products should be left on for 2 to 3 minutes, then rinsed off under warm running water, which will wash away the hair as well.

At this point, the mouse may be disinfected. The disinfectant also serves to wash off any loose hairs and mats down hair on an unshaven mouse for ease of examination. After letting some of the disinfectant drain from the mouse back into the container, place the mouse on one to two layers of absorbent paper towel on a cork board. The cork board should be approximately 14 cm × 21.5 cm in size, which is large enough for two mice, yet easy to move about during the necropsy. It should be at least 1.0 cm thick.

If skin is to be collected for characterization of a mutant, it should be done at this point. With the mouse ventral surface down on the board, gently grasp a fold of dorsal skin from the caudal region and make a small incision with the scissors. Carefully cut out a rectangular piece of skin along the dorsal midline from the thoracolumbar junction to the interscapular region and collect it in the desired medium. We often collect skin from one mouse for many different assays. Preservation methods are discussed in the fixatives section, but it is important to remember that presentation is just as important as your choice of fixative. Skin should be laid out flat on a piece of aluminum foil or mesh backing for fixation so that it may be cut neatly into thin strips for histology. Orient the skin sample cranial–caudally, and trim lengthwise on this axis to optimize the orientation of the hair follicles.

After collecting dorsal skin, collect a sample of ventral skin in a similar fashion. Fresh samples of both dorsal and ventral skin may be frozen in OCT for antigen expression studies. The skin from the head should also be collected, including the pinnae (ears), eyelids, and muzzle. All may be carefully peeled and trimmed from the skull as a unit and mounted flat on a piece of aluminum foil for fixation. Tail skin may be collected by severing the tail from the body and making an incision with a tip of a scissors down its length. Grasp the loose skin from the base of the tail, and strip the skin away from the bone and tendons. Mount the tail skin flat on a piece of foil as you did the other samples. Tail and head skin are often collected toward the end of the necropsy.

After skin has been collected, place the mouse ventral side up and pin each limb firmly to the board. The rear feet may be pinned between the gastrocnemius tendon and the bone, while the front feet may be pierced through the skin, between the metacarpal bones, in order to do the least damage to the tissues being collected. During the necropsy, the board may be rotated easily to adjust the position of the mouse, providing different angles of access for organ collection.

If the skin is not to be collected for a study, you will begin your necropsy by pinning the mouse down, ventral side up, as describe above. With either a #12 scalpel blade and #3 handle or a pair of sharp/sharp iris scissors, make a ventral midline longitudinal incision through the skin, from the external genitalia to the ramus of the mandible, then cut from the genitalia laterally toward the rear feet, along the medial surface of the rear legs. On female mice, this incision passes between the fourth and fifth nipples of each side. Grip the skin on either side of the incision and pull gently outward, or use light strokes of the scalpel, to separate the skin from the abdominal muscles. Reflect the skin far enough so that it does not interfere with the rest of the necropsy.

With the skin reflected back, collect the external lymph nodes. The peripheral lymph nodes are located on either side of the salivary gland (cervical lymph nodes), under each of the fore legs (axial lymph nodes), and on the inside of each rear leg (inguinal lymph nodes). The cervical lymph nodes are collected attached to the salivary gland. The inguinal lymph nodes can be located, if you look carefully, in the fat pad on the medial surface of the rear leg. They may be slightly darker in color than the fat and are usually no larger than 0.10 to 0.20 cm. The easiest way to collect them is to remove the fat pad with the lymph node embedded in it and trim away as much of the fat as possible before placing the lymph node in the fixative.

Next, using a clean, sterile set of instruments, grip the abdominal muscles at the inguinal region with forceps and lift firmly, making a small incision to let air into the abdomen. This will cause the abdominal wall to stretch further and the viscera to fall away so a more aggressive incision can be made without contaminating the tissues inside. Continue the cut through the abdominal muscles on each side, extending from the inguinal midline to the lateral thorax. This will neatly expose the viscera. If culturing from swabs within the abdomen is necessary, a third sterile set of instruments should be used at this time, with a new set used for each culture taken.

After any necessary abdominal samples have been taken for microbiological culture, tissue collection may begin. As a general rule, no more than fifteen minutes should pass between euthanization of the healthy mouse and collection of its tissues. This should allow adequate time for photography, culturing and other procedures. Throughout the necropsy, the visceral organs should be evaluated to determine if they are in their proper anatomic orientation. Some mutations such as situs inversus (*iv*) can cause the orientation to be reversed.[19] All tissues and organs should be carefully checked for abnormalities and all observations should be noted. Gross photographs of any external or internal abnormalities are important to obtain in the study of a new mutant. See Chapter 6, Photography of Laboratory Mice, for details.

The intestines should be collected first because autolysis begins quickly in these organs. Grasp the cecum with forceps, lift it up and cut away the small intestine at the rotund sac, where the ileum meets the cecum. This is the smaller of the two intestines joining at the cecum. Drop the cecum and grasp the severed end of the ileum, pulling it gently as the small intestine begins to unravel. Carefully cut any mesenteric tissue that provides resistance. Be sure not to puncture or sever the intestine itself, as this will make it more difficult to inflate and prepare for histology. If the intestine does break at any point, simply lay the broken section on a moist portion of the paper towel, maintaining proper orientation, then continue the procedure. Toward the duodenojejunal flexure, the pancreas is firmly adherent to both the small and large intestine. Care must be taken to separate the two intestines and the pancreas, while preserving the integrity of all three. Continue freeing the small intestine until you reach the stomach, then release the tension on the intestine. Before separating these organs, open the small intestine just distal to where it joins the stomach. Gently compress the gall bladder. Bile should empty out of the common bile duct, indicating that the duct is patent. Now cut away the stomach from the small intestine at the pars pylorica. Lay the intestine down in a large loop around the mouse remembering which end is the duodenum and which is the ileum. Intestinal rolls must be oriented the same each time to allow the pathologist to interpret them properly. It helps to develop habitual patterns of tissue collection to avoid errors.

For proper fixation, the intestine must be inflated with fixative before it is rolled or otherwise prepared for histological presentation. This procedure is simple and rapid but runs the risk of injury by squirting fixative into eyes accidentally. Be sure to take precautions, wear safety goggles, and use a fume hood if one is available. Use a 10 cc syringe with a 17 to 22 gauge needle. Fill the syringe with fixative and introduce the needle into either end of the intestine. Gently depress the plunger on the syringe and the intestine will slowly inflate. Making several injections along the length of the intestine is generally safer for the technician and is often necessary to provide proper inflation (Figure 5.4). An intestine that has been over inflated is much more difficult to roll. To make a second injection, simply pierce the wall of the intestine and clamp it firmly around the tip of the needle to prevent backflow of the fixative and proceed as before.[10] Some laboratories prefer to open the entire intestine and remove digested food material and feces prior to rolling.[20] This approach yields good mucosal fixation but may also cause damage to the mucosa.

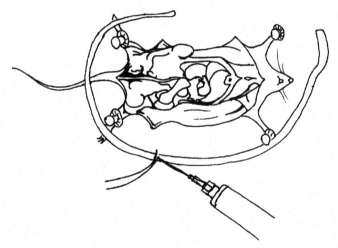

**FIGURE 5.4** Intestines are removed and inflated by injection with fixative. (Drawn by Ingrid K. Sundberg.)

Intestines can be rolled full length for histological presentation.[10,20] Each roll must be able to fit comfortably in a histological cassette. In an average sized adult mouse, this requires the small intestine to be cut into three equal pieces. With practice you will learn to judge the number of sections required for proper sized rolls.

Another approach to presenting intestines is similar to the way they are routinely collected in larger animals. Representative segments are cut and fixed by immersion. More precision comes by laying out the entire gastrointestinal tract and cutting segments out at specified distances from the anus or pyloris.

To create what are called "Swiss Rolls" for full-length histological presentation, roll the inflated intestine in concentric, centrifugal circles on a piece of unlined index card (Figure 5.5), letting as little fixative as possible flow out of the segment. If the fixative drains, the intestine may flatten, making it more difficult to roll. We use large index cards cut into strips approximately the width of our histological cassettes to mount the tissues on. As mentioned previously, the orientation of the segments is important and must be agreed upon with the pathologist. It is commonly accepted to keep the end of the intestine proximal to the stomach toward the center of the roll. Once the intestines have been rolled onto the paper, let them stiffen for a few minutes before placing them into a jar of fixative, to prevent the roll from unwinding.

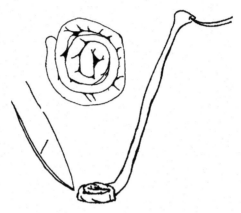

**FIGURE 5.5** "Swiss roll" of intestines. (Drawn by Ingrid K. Sundberg.)

Intestinal rolls take practice to master, and one may wish to develop his or her own techniques for the best presentation. The important objectives to remember are (1) do not over inflate the intestine, (2) keep the proper orientation, (3) make the rolls smaller than the cassettes, and (4) try not to let the fixative drain while rolling the segments.

Before the colon and attached cecum may be dealt with, one must first remove the reproductive organs, including the preputial gland or the clitoral gland, and urinary bladder. In the male mouse, testes may be gently grasped by the inguinal fat pad, cut away from the other viscera and placed on a bit of cardstock. Orient them in such a way that the epididymis and testis are in the same plane and may be trimmed simultaneously for histological presentation. The preputial gland is a paired organ located subcutaneously between the penis and the rectal opening. It may be collected by grasping one edge of the gland with your forceps, cutting it away from the abdominal wall and placing it directly into fixative. The seminal vesicles, urinary bladder, and penis may be removed as a unit by gently grasping the apex of the seminal vesicles with forceps and lifting it away from the colon. Insert the tip of heavy duty scissors between the colon and the pelvis and cut the bone on both sides, then lift the reproductive tract further and cut away the connective tissue between it and the colon. Remove the reproductive tract and arrange it on a card, then place it in the fixative.

The female reproductive tract is removed in a similar fashion. The clitoral glands are normally not easily visible, but are located subcutaneously just in front of the vaginal opening. The best way to collect these glands is to cut a small square of 0.5

cm to 0.8 cm of abdominal muscle and overlying skin immediately anterior to the clitoris. This section will include the clitoral glands. Smooth this piece of tissue gently onto a piece of white cardboard and, using parallel pencil marks, indicate the area where the glands are expected to be. After placing the clitoral glands in fixative, grasp the fat pad of one ovary and cut it free from the mesentery, laying that ovary and uterine horn over on the other side of the colon. Then cut the pelvic bone as before, cutting the other half of the uterus and its ovary free from the surrounding fat and moving it out of the way before severing the contralateral pelvic bone. Remove the entire female reproductive tract as you did the male organs, then arrange it on a card before fixation to maintain orientation.

Now grasp the cecum with the forceps and pull it gently, cutting away the mesentery to free the colon. Trim the anus away from the surrounding skin and lay the structure down on your moist work surface. Inflate the colon with fixative as you did the small intestine (Figure 5.6). Insert the needle carefully into the anus and slowly depress the plunger of the syringe. Fecal matter in the colon can block the flow of the fixative. Gently express the feces and then inflate with fixative. The cecum should be inflated just slightly through the ileocecal junction, cut from the colon, then laid out on a flat surface to stiffen before dropping it into the fixative. The colon should be rolled similar to the rest of the intestine, with the same orientation. The colon is often stiff and more difficult to handle due to the presence of fecal pellets, and should be left out to harden for quite some time once it has been inflated and arranged properly on the card.

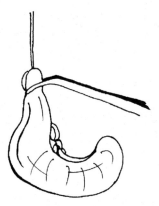

**FIGURE 5.6** Inflation of the cecum with fixative. (Drawn by Ingrid K. Sundberg.)

The last of the gastrointestinal tract to collect is the stomach. Grasp this organ at the pylorus and gently pull it away from the liver until you can see the esophagus, which presents itself as a narrow whitish structure extending from the center of the stomach upward. Cut this carefully while maintaining your grip on the pylorus and pull the stomach away from the liver, trimming away any mesentery or bits of pancreas that may adhere to the gastric surface. Inject a small amount of fixative into the stomach through the pylorus, similar to the way the cecum was inflated, then place it in the jar with the other organs.

The spleen and pancreas are connected and are often removed as a unit unless it is necessary to examine one or the other separately, as in the case of a mouse model for diabetes where special fixative (Bouin's solution) and stains (Aldehyde Fuchsin) are used to examine the beta cells of the pancreatic islets. Grasp the pancreas gently with the forceps and pull upward, cutting away any mesenteric tissue that adheres to the spleen or pancreas to free the structures, then place them directly into the fixative.

Kidneys may be removed by grasping the surrounding fat and pulling upward while cutting around the organs. The adrenal gland is a small white structure that lies within the perineal fat pad just anterior to the kidneys. It should be left within the fat that clings to the kidney and the two should be presented to histology as a unit. It is important for the pathologist to be able to distinguish between the right and left kidneys. When it comes time to trim the kidneys after fixation, the right kidney should be cut transversely, while the left is cut lengthwise (left/long) for histological presentation. A small transverse incision in the right kidney before fixation will allow you to distinguish between the two and improve fixation if left intact.

The liver is the last organ to be removed from the abdominal cavity. To accomplish this, it helps to enter the thorax so you may use the diaphragm as a handle to trim out the liver. Move the liver gently out of the way with the edge of a forceps. Then grasp the xiphoid process firmly with the forceps and pull upward to create a negative pressure within the thorax. Cautiously cut through the ribs and diaphragm of one side. This will allow the air to enter the thoracic cavity, and the lungs within will shrink away from the diaphragm. If a thoracic microbiological culture is to be taken, extend the cut through the diaphragm and rib cage to expose the lungs, then use sterile scissors and forceps to take a tissue sample before proceeding. After all samples for microbiological cultures have been taken, grasp the diaphragm with forceps and trim it completely away from the ribs, being cautious near the spinal column so as not to cut the liver. The liver may then be easily lifted away. Place the liver on a damp work surface and separate the right and left medial lobes as a unit, along with the gall bladder. This may be accomplished by folding these lobes back onto the work surface and trimming the ligaments that connect them to the remaining liver lobes. The lateral left lobe is the largest and should be separated from all others in a similar fashion. The remaining smaller lobes may be placed in the fixative together. When separating the liver lobes, take care to handle them gently. The liver is very fragile and is easily damaged.

The thoracic cut may now be extended through the rest of the ribs at the costochondral junction to just short of the internal thoracic vein and artery on both sides. Often the mediastinum between the area of the heart and thymus continues pulling on the excised part of the rib cage, so it must be carefully trimmed away to prevent the rib cage from falling back over the rest of your work.

At some point, before you cut into the cervical area, the salivary glands should be removed. These glands lie subcutaneously in the ventral cervical area. Gently pulling back the skin in this area will sufficiently expose the glands for removal. The tip of this lobed structure is narrower and lies closer to the thorax. Grasp it gently with your forceps and slowly pull upward, freeing it from the surrounding tissue with your scissors, then cut the glands horizontally across their base and place them in the fixative.

To remove the heart and lung, turn the cork board around so that the mouse's head is facing you. With your forceps, lift the lower jaw, (remember, the mouse is on its back) and cut through the hinge of the mandible with the scissors, separating the lower jaw from the rest of the skull. The epiglottis will now become apparent. Being careful not to puncture the trachea or the esophagus, continue your cut on either side of the neck until you reach the first rib. Using the tip of your scissors, sever the first rib and continue to dissect carefully through the remaining ribs, avoiding the trachea, esophagus, and lung. Repeat this on the other side. Once all the surrounding tissue has been removed, pull up gently on the mandible, trimming any remaining mesenteric tissue from the underside of the lungs. The entire structure, including the lungs, heart, trachea, epiglottis, lower jaw, and tongue will be removed as one. The lungs are another structure that must be inflated with fixative to ensure proper histological presentation. Use the same syringe and needle setup as you used for the intestines. Slip the tip of a needle into the trachea via the glottis, which is normally the most apparent hole at the base of the tongue. Clamp down around the needle with a pair of forceps and slowly depress the plunger of the syringe. If the lungs begin to expand and blanch, you have successfully found the trachea. If the fixative flows out in a puddle between the lungs, try again. Inflate the lungs very slowly, stopping when they are about the size that they would normally be on inhalation. Overinflating the lungs can damage the alveoli, often causing diagnostic difficulty.

Once the lungs have been inflated, cut through the trachea and esophagus to separate the mandible from the heart and lungs. The heart, lungs, and thymus should remain together as a unit for fixation. The tongue should be separated from the mandible by grasping the tip of the tongue with your forceps and cutting it off at the base. Place the tongue, mandible, heart, lung, and thymus into the fixative.

The next organ to be collected is the brain. To access the brain, cut the spinal cord at the base of the skull. Slowly pull and cut the skin away from the skull if the skin has not already been removed. Be sure to cut carefully around the eyes, leaving the eyelids attached to skin and avoiding damage to the eyes themselves. To focus on the eyes, you may wish to remove them from the skull. To remove the eye, sink a pair of curved forceps behind the orbit, grasp the optic nerve and pull outward until the eye has been freed from the socket and may now be fixed and embedded separately. However, histological presentation of the eye within the skull is often sufficient for viewing many abnormalities.

Cut any remaining vertebrae off the skull. You will see a mass of white matter protruding from the foramen magnum. This is the spinal medulla. Slip one blade of your scissors in between the neural tissue and the bone and make two small longitudinal cuts in the occipital bone (Figure 5.7), one on each side of the spinal medulla. Hold the skull between your thumb and forefinger and, with your forceps, grasp the edge of the occipital bone and pull upward to neatly break it off. Then, gently insert the tip of your scissors between the brain and the skull and make a cut in the skull along the sagittal suture (Figure 5.8). Continue breaking away the interparietal and the parietal bones in the same manner, being careful not to harm the delicate brain below. The frontal bone comes to a slight point at the intersection of the sagittal and coronal sutures. This area should be broken off as well, to allow the brain to be removed cleanly from the cranial cavity. There may be a thin reddish membrane around the brain, particularly in the area of the cerebrum. This is the meninges and must be removed carefully with forceps or it will cut into the brain as you try to remove the brain from the skull. Once you have removed the meninges, turn the skull upside down over the jar of fixative. Gently work the forceps between brain and bone and pull away the connective tissues, freeing the brain from the cranial cavity. The brain will fall into the fixative. Place the skull into the fixative as well.

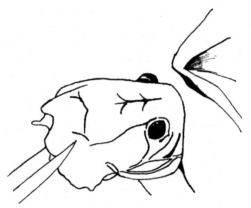

**FIGURE 5.7** Initial cuts in the occipital bone to access the brain. (Drawn by Ingrid K. Sundberg.)

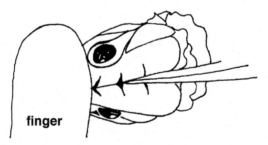

**FIGURE 5.8** Second set of cuts down the sagittal suture for exposure of the brain. (Drawn by Ingrid K. Sundberg.)

The spinal column may now be collected. Grasp the proximal end of the spine and lift it away from the skin, cutting away the fascia that hold the two together. Cut through the pelvis to sever the hind limbs from the distal vertebrae, then cut the tail away as well. Cut the ribs away from the spine, as close as possible, without damaging the vertebrae. Place the spinal column in fixative, making sure to keep it straight so it is oriented correctly when it is time to trim the tissues.

Trim away the front and rear limbs from the remaining skin and place them in fixative. If skin was not collected earlier in the necropsy, it may still be important to save it for your archives. We take the full pelt of the mouse and fold it in half over a strip of cardstock before placing it in fixative. This is a quick and easy way to save the skin in an orderly fashion for future use, if necessary.

In summary, when working up the effects of a new mutation, it is important to collect study sets of total tissues in a methodical, standardized fashion to avoid diagnostic discrepancies due to presentation. From these study sets, a more focused tissue collection protocol may be developed, concentrating on those tissues known to express abnormalities in your mutant. Always remember the importance of adopting standard criteria for tissue collection, agreed upon by the technicians, researchers, and the pathologist who will be reading the slides.

## X. TRIMMING TISSUES FOR HISTOLOGY

After the tissues you have collected have been fixed for a sufficient amount of time (see section on fixatives for different guidelines), they must be trimmed before being delivered to the histology laboratory for embedding. Proper trimming of the tissues will ensure that they are presented on the slide in an orientation that will allow appropriate interpretation of their cellular structure by the pathologist. Presentation is critical when trying to identify any variation from "normal" or any pathologic change present in the tissue being viewed.

Before the tissues can be trimmed, it is necessary to decalcify the bones. This is accomplished by an overnight soak in a hydrochloric acid decalcifying solution such as Cal-EX (Fisher, Pittsburgh, PA). After decalcification, bones may be immediately trimmed after an initial rinse with water, but must be continually rinsed in running water for at least 3 or 4 hours before being processed by the histology laboratory. Failure to thoroughly rinse decalcified tissues may result in inadequate staining of the tissue sections. Optimization of the times required for decalcification and washing should be customized for every laboratory.

After any necessary decalcification has been completed, trimming of all tissues may commence. Use a cork board, similar to the one used during your necropsy, as the base on which to trim your tissues. Each tissue, as it is trimmed, should be placed into a labeled and numbered histology cassette (OmniSette Tissue Cassettes, Fisher Scientific, Pittsburgh, PA). Use a #2 pencil, solvent-resistant marker (Histo-Prep Pen, Fisher Scientific), or mechanical labeling machine to label the front and/or side of each cassette. Indicate the mouse's accession number and, if there will be more than one cassette of tissue per mouse, number each cassette for that animal in sequential order. Sequential numbering will assist you in determining whether all tissues sent to histology are returned. Furthermore, knowing which tissue went into which cassette will help identify specific tissues (left vs. right, liver vs. kidney, tumor 1 vs. tumor 2, etc.). Some laboratories always put the same tissues into the same numbered cassette for standardization. Cassettes containing tissues that were fixed and not decalcified should be placed into a container of 70% ethanol, while those containing decalcified tissues not previously rinsed should be placed in water. After a thorough rinsing, decalcified tissues may be placed into ethanol with the other tissues and delivered to the histology laboratory for processing and embedding. Any remaining tissues not submitted for embedding may be stored in 70% ethanol for future use, if necessary.

When handling fixed tissues, it is still important to be gentle. Soft parenchymal organs, such as the liver, brain, lungs, kidneys, etc., remain delicate after fixation and should be manipulated using a pair of wide wooden forceps or a similar tool. Trimming is best done with a single-edge razor blade, which, if sharp, can allow a clean, concise cut with minimal damage to the tissue. Residual chemicals on the tissues can dull a razor blade rapidly, so it is important to change to a new blade frequently while trimming.

All tissues should be trimmed to a thickness no greater than the depth of the histology cassette (4 to 5 mm). If it is necessary to present a particular facet of a trimmed tissue to the pathologist, this must be indicated by marking the side of the tissue one does *not* want presented. A blue colored pencil works well (Venus col-erase, #1276 blue, Eberhard Faber, Inc., Lewisburg, TN), because the blue pigment will not wash off in alcohol and will clearly indicate the desired orientation to the histologist who will be embedding the tissue. Be sure to mark the side not to be sectioned, because the blue pigment will contaminate the field in a histological section, making it difficult for a pathologist to correctly interpret the tissue being presented. Small, related tissues of similar densities may be placed together in the histocassettes (i.e., kidneys with spleen and/or liver, reproductive tract tissues together, etc.). The heart, lungs, brain, bones and any possible tumors found should each go into a cassette by themselves. The following is a description of trimming technique for each organ. As mentioned previously, any remaining tissue not sub-mitted for embedding may be saved for future use in 70% ethanol.

*Large and small intestine*—Intestines need no extra trimming at this point, because they should have been prepared before fixation. Each segment should be carefully removed from its backing, if rolled, and placed individually in a histocas-sette.

*Stomach and cecum*—Both these organs should be cut in half longitudinally. The stomach should be cut to present both the esophageal and duodenal openings. The cecum should be cut to show both the ileocecal junction and the ampula of the colon. Submit the half of each that best shows the desired features. If space allows, the cecum and stomach may be submitted in the same histocassette. If your study is focused on lesions of the cecum or stomach, it may be important to submit both halves of one or both of these organs. If this is the case, each organ should be placed in its own cassette.

*Liver (with gallbladder)*—Cross sections of only the left lateral lobe of the liver and the medial lobe with the gall bladder may be sent to the histology laboratory unless any pathologic changes are obvious only on the accessory lobes. Lay the left lateral lobe out flat on your trimming surface and cut a crosswise section from anywhere near the center of the lobe (Figure 5.9). Trimming the medial lobes must be a bit more precise, because they must be cross-sectioned in such a way as to include a portion of the gall bladder in the section. This is usually accomplished by cutting across the medial lobes just below the juncture where the lobes separate, then just above that juncture, where the falciform and teres ligaments hold the two lobes together. The first cut should reveal a portion of the gall bladder. Mark the opposing side with blue pencil and place both this section and the left lateral lobe section in the histocassette.

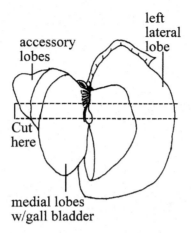

**FIGURE 5.9** Example of trimming sites for a mouse liver. (Drawn by Ingrid K. Sundberg.)

*Kidneys*—The left kidney should be cut longitudinally down the center and should include a segment of the adrenal gland which was left attached to the kidney at the time of fixation (Figure 5.10). The right kidney will be identified by a small transverse incision, if handled properly at the time of fixation. It should be presented in a lateral cross section cut through the central area near the pelvis. Use the blue pencil to mark the side of this section farthest from the pelvis, because it is important to present the area nearest the center of the kidney to the pathologist. The kidney sections may be placed in the histocassette with the liver sections.

*Spleen and pancreas*—Unless you have collected the pancreas separately for focused study, the spleen and pancreas should have been fixed as a unit. Trim them together in cross section at any point along the length of the spleen. This section of the spleen and pancreas may be submitted in the same cassette as the liver and kidneys, if space allows.

*Lungs*—The lungs are collected with the heart and thymus as a unit at the time of necropsy but are submitted separately for histological studies. Using a pair of wooden forceps, gently push apart the heart and lungs so that the lungs lay out flat. Cut a longitudinal section from the center of the lobes of the lung on each side. The individual lobes will separate, but should all be collected and placed in the same histological cassette. There is generally no need to separately identify each lobe of the lung.

*Heart and thymus*—Carefully remove any remaining lung or tracheal tissue from around the heart, being certain not to separate the thymus from the heart. Place the heart on its base and begin your cut at the heart's apex. Angle your cut so that it bisects each of the four chambers of the heart (Figure 5.11). In some instances, the coronary artery may be visible on the epicardium of the fixed heart. Making a cut along the line of this vein will often bisect the chambers properly. It may take time to become familiar with the external features of the heart before you can regularly make this cut correctly. If space allows, both halves of the heart may be submitted. If only one half is submitted, make sure that it contains a portion of the thymus.

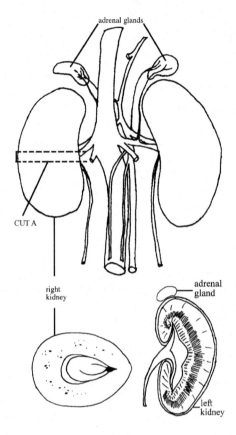

**FIGURE 5.10** Anatomic location of both kidneys and adrenal glands. Right kidney is trimmed transversely, while left is cut longitudinally so that histologic sections can be identified. Lower panels illustrate features of cut sections. (Drawn by Ingrid K. Sundberg.)

**FIGURE 5.11** The heart is trimmed lengthwise to include all four chambers. (Drawn by Ingrid K. Sundberg.)

*Salivary gland*—Trim the base of the salivary gland to present a clean-cut edge. Make your second cut approximately 4 mm from the first cut. Mark the face of the second cut with blue pencil and place the cross section in a histocassette.

*Trachea and thyroid/parathyroid*—Cut a cross section of the trachea at the point where the thyroid and parathyroid glands are attached. This is located in an area 1 to 2 millimeters below the epiglottis. A V-shaped cut made between the molars at the base of the tongue will free the soft tissue from the bone, or the entire lower jaw may be decalcified. If decalcified, the cut to reveal the trachea and surrounding glands may be made straight across the base of the mandible so that the section includes the molars and bones of the mandible.

*Lymph nodes*—There are many lymph nodes located throughout the body, but it may not be necessary to submit all of these for processing unless involved in the disease process. Representative nodes from several areas may be chosen (i.e., Mesenteric, axial, inguinal, cervical). If not enlarged, lymph nodes may be submitted whole, after the surrounding fat has been removed. Severely enlarged lymph nodes must be cut in cross section.

*Urinary bladder*—The urinary bladder should be fixed as a unit with the reproductive organs. A lengthwise cut made down the center of the urinary bladder should also include the uterine body, vagina, and cervix in the female, and the penis and prepuce in the male (Figure 5.13). After this first cut is made, trim the opposite side of one half of your tissue to the appropriate width to fit in your cassette (~4 to 5 mm), cutting off the uterine horn or seminal vesicle and any excess adipose tissue. Mark this side with blue pencil and place the section in a histocassette.

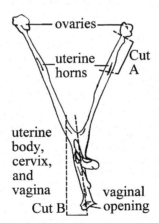

**FIGURE 5.12** Female reproductive tract. Dotted lines indicate where to cut tissue for histologic processing. (Drawn by Ingrid K. Sundberg.)

*Reproductive organs*—Female: Cut one ovary away from either uterine horn, trim away excess fat and place the entire ovary and its associated uterine tube in a histocassette (Figure 5.12). Into the same cassette, place a cross section of one of the uterine horns. Trimming the uterine body, vagina, and cervix is discussed above along with the urinary bladder.

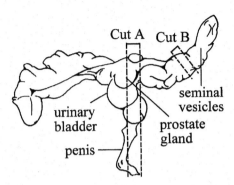

**FIGURE 5.13** Male reproductive tract. Dotted lines indicate sampling sites for histology. (Drawn by Ingrid K. Sundberg.)

Male: Separate the testes from the index card on which they were fixed. The testes and epididymis should be slightly flattened and lie in the same plane on the surface that was attached to the card. Mark the rounded side of the testis with blue pencil and place into a histocassette. Into the same cassette, place a cross section from one side of the seminal vesicles (Figure 5.13). The penis and prepuce are discussed above along with the urinary bladder.

In both male and female, all reproductive organs, as well as the urinary bladder, may be submitted in the same cassette.

*Clitoral/preputial glands*—The clitoral gland of the female mouse is normally quite small and is fixed in a segment of inguinal fascia and fat. This segment should be cut in cross section and embedded on edge. It may be submitted in the same cassette as the reproductive organs. The preputial gland of the male mouse is larger and is collected individually. It should be cut in cross section as well and sent with the reproductive organs.

*Brain*—The brain is submitted to the histology laboratory cut in three cross sections rostral to caudal (Figure 5.14). The first cut should be made through the cerebrum, ~1 to 2 mm from the olfactory lobes. The second cut through the cerebrum, 2 to 4 mm from the first, will create the first section. This first section should present a view of the central portion of both cerebral hemispheres, so mark the face of the *first* cut with blue pencil. The third cut should be made 2 to 4 mm from the second cut, just to the cerebral side of the confluence of sinuses. This will create the second cross section, of which the face created by the *second* cut should be marked with blue pencil. A fourth cut should be made at the transverse sinus to separate the remaining portions of the cerebral hemispheres from the cerebellum, after which a fifth cut is made, which approximately bisects the cerebellum laterally to create the third section. On this section, the cut face closest to the cerebrum should be marked with blue pencil. The brain should be submitted in a cassette by itself.

*Tongue*—The tongue should be cut longitudinally down the midline. It is only necessary to submit half of the tongue to histology.

*Legs (long bones)*—One each of the fore and hind legs should be cut longitudinally to show the long bones and major joints of each leg (Figures 5.15 and 5.16). Excess fat should be trimmed away and on the hind legs it may be necessary to trim

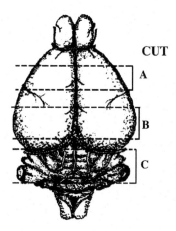

**FIGURE 5.14** Trimming sites marked for sectioning a mouse brain. (Drawn by Ingrid K. Sundberg.)

away some of the bulk of the muscle in order to fit the section properly into the cassette. Feet should also be separated from the leg, and will be submitted separately. The longitudinal cut should be made using the major joint of each leg as a reference to bisect the long bones. Choose the half of each leg that best shows the desired view of the bones and place it in its own histocassette.

*Feet*—One each of the front and back feet should be cut longitudinally to show the skin, foot pads and bones of the feet. The front foot should be cut so two toes are present on each half. The back foot should be cut directly through the middle toe on that foot. Each foot is placed into a separate histocassette. Both halves of each foot may be sent to the histology laboratory.

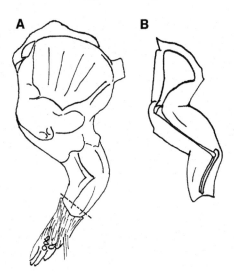

**FIGURE 5.15** Front leg. (A) Trim site with skin removed. (B) Decalcified limb cut lengthwise to expose joints. (Drawn by Ingrid K. Sundberg.)

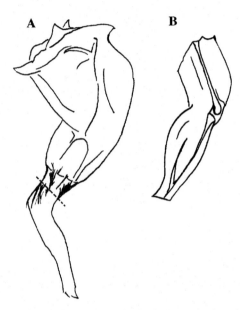

**FIGURE 5.16** Rear leg. (A) Preliminary preparation of leg includes removal of skin and amputation of distal segment as indicated. (B) Decalcified limb is cut lengthwise to expose joints. (Drawn by Ingrid K. Sundberg.)

*Spinal column*—The spinal column should be trimmed to present both lateral and longitudinal sections of the thoracic and lumbar regions. First, cut the spine laterally between the thirteenth thoracic and first lumbar vertebrae (just below the 13th rib). Next, cut a lateral section 4 to 5 mm wide, from the distal end of each section. Bisect the remaining long segments longitudinally, placing the best half of each into a separate cassette with its related lateral section (Figure 5.17).

**FIGURE 5.17** Longitudinal section of decalcified spinal column. (Drawn by Ingrid K. Sundberg.)

*Skull*—The skull should be cut into three cross sections, similar to the brain (Figure 5.18). The sections should show the eyes, nasal passages, ear canals and pituitary gland. The first cut should be made through the posterior edge of the pituitary, identified as a whitish mass located in the approximate area between the occipital bone and basiphenoid bone on the inner surface of the skull. This is the crucial cut and should present a view both of the pituitary and the middle ear. The second cut may be made 4 to 5 mm anterior to the first cut to create the first section. The third cut should be made through the posterior edge of the visible portion of the eyes, with the following cut made just anterior to the eyes. The third section should present a view of the sinuses, and may be cut from the approximate center of the remaining portion of the snout. Each section should be marked with blue pencil on its anterior surface. All sections of the skull may be submitted in the same cassette.

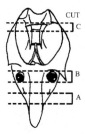

**FIGURE 5.18** Three sections are cut in the decalcified skull. This exposes: (A) the nasal cavity, (B) eyes and associated glands, and (C) inner, middle, and external ear, as well as the pituitary gland. (Drawn by Ingrid K. Sundberg.)

*Skin*—Trim portions of the dorsal and ventral skin longitudinally in the direction of the hair growth, into pieces approximately 0.3 × 1.0 cm. Cut 2 to 3 sections of both dorsal and ventral skin in this manner and mark one long edge with blue pencil to indicate that the pieces should be embedded on the opposite edge. Depending on the focus of your study, you may also want to cut a piece of skin approximately 0.7 × 0.7 cm square to be submitted horizontally, haired side down, to view the hair follicles in horizontal section. Mark the underside of the section with blue pencil. Tail skin should be trimmed in the same orientation as the longer pieces of dorsal and ventral skin. Cut the section from an area that was not handled as the tail skin was removed from the bone, and mark one long edge with blue pencil. Eyelids may be presented by making a cut bisecting the lids of both eyes, then making a second cut just posterior or anterior to the corners of the eyelids so that the upper and lower eyelids remain attached to one another (Figure 5.19). Mark the face of the second cut with blue pencil. A section of the muzzle skin may be obtained by making a cut approximately 3 to 4 mm in from the front edge of the muzzle (Figure 5.19). Mark the outer, uncut, edge with blue pencil. Trim a section out of one ear by first cutting one of the ears in half lengthwise and then cutting one of the halves completely off from the scalp (Figure 5.19). Lay this half flat on your cutting surface and make a second cut parallel to the first to obtain a section similar in size and shape to that of the dorsal and ventral skin sections. Put a blue mark on one of the long sides of the section of ear. The various skin sections may be combined into histocassettes. However, you should combine the same pieces every time, so if you have questions later about whether you are looking at dorsal or ventral skin you can figure things out by looking at the other types of skin included. We combine the long sections of dorsal skin with the sections of ear and tail skin into one cassette. The long sections of ventral skin go with the sections of eyelid and muzzle in a second cassette, and we put each of the square sections of dorsal and ventral skin into individual cassettes. Each cassette is also labeled with a "D" or "V" to aid in identification.

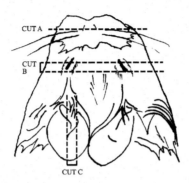

**FIGURE 5.19** Skin of the head trimmed to study (A) muzzle and vibrissae, (B) eyelids, cilia, Meibomian gland, and conjunctiva, and (C) pinna of the ear. (Drawn by Ingrid K. Sundberg.)

Hematoxylin & Eosin (H&E) stain should be requested on all tissues sent to the histology laboratory. Special stains may be requested on specific tissues if this is necessary to verify any suspected pathologic changes not sufficiently disclosed by the H&E stain. Protocols for various stains, what they stain, and what colors they stain are subjects of various histology and pathology text books.[17,21]

## XI. CONCLUSIONS

A research-quality necropsy and interpretation of gross and microscopic changes requires a great deal of skill and anatomical knowledge on the part of both the technician and the pathologist. Both need to work together to coordinate efforts and optimize protocols to achieve consistent, high-quality results. This chapter provides an overview on how to achieve these results, but practice is required to develop the skills.

## ACKNOWLEDGMENTS

This work was supported by grants from the National Institutes of Health (CA34196, AR43801, and RR8911) and the Council for Nail Research.

## REFERENCES

1. *Stedman's Medical Dictionary,* 26th ed., Spraycar, M., Ed., Williams & Wilkins, Baltimore, MD, 1995.
2. Sundberg, J.P., Boggess, D., Sundberg, B.A., Beamer, W.G., and Shultz, L.D., Epidermal dendritic cell populations in the flaky skin mutant mouse, *Immunol. Invest.*, 22, 389, 1993.

3. Sundberg, J.P., Rourke, M., Boggess, D., Hogan, M.E., Roop, D.R., and Bertolino, A., Angora mouse mutation: altered hair cycle, follicular dystrophy, phenotypic maintenance of skin grafts, and changes in keratin expression, *Vet. Pathol.*, 34, 171, 1997.

4. Sundberg, J.P., Boggess, D., Hogan, M.E., Sundberg, B.A., Rourke, M.H., Harris, B., Johnson, K., and Davisson, M.T., Animal Model. Harlequin ichthyosis (*ichq*). A juvenile lethal mouse mutation with ichthyosiform dermatitis. *Am. J. Pathol.*, 151, 293, 1997.

5. Mohr, U., Dungworth, D.L., Capen, C.C., Carlton, W.W., Sundberg, J.P., and Ward, J.M., *Pathobiology of the Aging Mouse*, Vols 1 and 2, ILSI Press, Washington, D.C. 1996.

6. Sundberg, J.P., Beamer, W.G., Shultz, L.D., and Dunstan, R.W., Inherited mouse mutations as models of human adnexal, cornification, and papulosquamous dermatoses, *J. Invest. Dermatol.*, 95, 62S, 1990.

7. Nanney, L.B., Sundberg, J.P., and King, L.E., Increased epidermal growth factor receptor in *fsn/fsn* mice, *J. Invest. Dermatol.*, 106, 1169, 1996.

8. Percy, D.H. and Barthold, S.W., *Pathology of Laboratory Rodents and Rabbits*. Iowa State University Press, Ames, 1993.

9. Hall, C.S., Genetic differences in fatal audiogenic seizures, *J. Hered.*, 38, 3, 1947.

10. Sundberg, J.P., Elson, C.O., Bedigian, H., and Birkenmeier, E.H., Spontaneous, heritable colitis in a new substrate of C3H/HeJ mice., *Gastroenterology*, 107, 1726, 1994.

11. Kiupel, M., and Sundberg, J.P., Bulbourethral gland abnormalities in inbred laboratory mice, *Lab. Anim. Sci.* (submitted).

12. Mähler, M., Bedigian, H.G., Burgett, B.L., Bates, R.J., Hogan, M.E., and Sundberg, J.P., Comparison of four diagnostic methods for detection of *Helicobacter* species in laboratory mice, *Lab. Anim. Sci.*, 48:85-91, 1998.

13. Popesko, P., Rajtová, V., and Horák, J., *A Colour Atlas of Anatomy of Small Laboratory Animals*, Vol. 2, Wolfe Publishing Ltd., London, 1992.

14. Feldman, D.B. and Seely, J.C., *Necropsy Guide: Rodents and the Rabbit*, CRC Press, Inc., Boca Raton, 1988.

15. Fekete, E., A comparative morphologic study of the mammary gland in a high and low tumor strain of mice, *Am. J. Pathol.*, 14, 557, 1938.

16. Sundberg, J.P., Montagutelli X., and Boggess, D., Systematic approach to evaluation of mouse mutations with cutaneous appendage defects, *Molecular Basis of Epithelial Appendage Morphogenesis*, Chuong, C-M. Ed., R.G. Landes Co., Austin, TX, 1998, 422.

17. Luna, L.G., *Manual of Histologic Staining Methods of the Armed Forces Institute of Pathology*, McGraw-Hill, Inc., New York, 1960.

18. American Veterinary Medical Association, 1993 Report of the AVMA panel on euthanasia, *J. Am. Vet. Med. Assoc.*, 202:229-249, 1993.

19. Layton, W.M., Random determination of a developmental process: reversal of normal visceral asymmetry in the mouse, *J. Hered.*, 67, 336, 1976.

20. Moolenbeek, C. and Ruitenberg, E.J., The Swiss roll: a simple technique for histologic studies of the rodent intestine, *Lab. Anim.*, 15, 57, 1981.

21. Smith, A. and Bruton, J., *Color Atlas of Histological Staining Techniques*, Year Book Medical Publishers, Inc., Chicago, 1977.

# 6 Photography of Laboratory Mice

*John P. Sundberg and James Miller*

## CONTENTS

## I. INTRODUCTION

Photodocumentation is an important aspect of defining a new mouse mutant phenotype and comparing it with similar human diseases. It provides a permanent record of the findings, but more important, it serves as visual support for descriptions in the text. This is particularly important when molecular biologists, usually not fluent in medical terminology, inadequately describe the mutant mice under investigation, making it difficult for subsequent investigators to understand the original observations and interpretations.

High-quality equipment is readily available or can be obtained through institutional audiovisual programs. Most pathology laboratories are well equipped to provide this support function. Once the equipment is set up and personnel are comfortable with its use, minor attention to details will yield excellent and reproducible results requiring very little time and effort.

This chapter will outline the equipment and how to set it up for photographing mice for publication.

## II. CAMERAS AND LENSES

A wide variety of cameras and lenses are available and the ultimate product decision depends upon individual preference. The first decision deals with the image recording format of the camera to be used. Current choices are conventional chemical-based films and processing vs. electronic imaging using either analog or digital technology. Chemical-based film and processing has three major advantages: (1) it produces the highest resolution image, (2) records are permanent, although they can be damaged if not properly handled, and (3) equipment and supplies are available and relatively inexpensive. Electronic image generation also has advantages and disadvantages. Analog technology is useful for teaching because it produces a real-time image that will not be interrupted by movement. Images can be captured on computers with specialized image capture boards and manipulated immediately for generation of final prints, integration into composite images, or image analysis. For still images, this technology is rapidly becoming replaced by digital imaging. Digital imaging has many of the same benefits, and resolution has reached the level necessary for many medical journals. This provides a rapid and less expensive way to publish in color. Many journals are moving in this direction, and more will follow in the years to come. The major disadvantage is that the equipment required to provide high-resolution images is very expensive, although this is changing. Chemical-based images (2 × 2 transparencies or prints) can be digitized at a later date, providing a suitable alternative for selected images. Scanners can be purchased with slide imaging capabilities for generation of analog images or slides can be sent to commercial service bureaus to be digitized and the files put on Photo-CDs that can be imported in a variety of formats. One big advantage of electronic imaging is that either black-and-white or color images can be produced from the same file.

Electronic imaging cameras are changing rapidly so any discussion of these would be out of date before this chapter appears. However, most red/green/blue (RGB)-type video camera can utilize a "C" mount to adapt lenses used by 35 mm lens reflex (SLR) cameras. These cameras can then be mounted on a suitable copy stand or tripod to be used in an identical manner as would a 35 mm SLR camera. Therefore, the general information on setups can be adapted for gross photography regardless of the camera back selected.

The classic camera system used for gross (whole animal) medical photography is a 35 mm SLR with a through-the-lens metering system. Ideally, two matching cameras and lenses are used, one for color film and the other for black-and-white film. The cameras do not need to be complicated or expensive. In fact, overly automated cameras with many features tend to be too complicated for routine work when many individuals will be using the equipment. The autofocus models appear to be useful but are generally not recommended for macro work, as the sensor may not focus where you want it. For these reasons, a manual-focus camera is more practical and also less expensive. Lens choice varies, depending on their ultimate use. A 50 or 100 mm macro lens is best for general use in mouse photography. This will be adequate to photograph the entire mouse and some larger structures. Extension tubes or a bellows unit, for the model of camera and lens used, are valuable for obtaining higher magnification (> 1X or 1:1), close up images of small lesions on individual organs. These are superior to screw-in type auxiliary lenses that attach to the front of the camera lens or $2 \times$ or $3 \times$ teleconverters that fit between the lens and camera body. Neither of these systems is optimized for the lens and camera. This reduces the quality of the final image.

## III. FILMS

Chemical-based films are available from a number of commercial producers, and each laboratory group will determine which it prefers. We use Kodak (Eastman Kodak, Rochester, NY) color slide (Ektachrome EPY 64T and EPT 160T) and black-and-white negative (TMX 100 and Tri-X Pan) films. Color slide films need to match the temperature of the lights used to illuminate the specimen. If tungsten lights are used, films labeled with a "T" are used without color correction filters. Daylight films will have a brown cast with tungsten lights or blue with fluorescent lights. Color correction filters can be used to eliminate these problems. These are not issues for black-and-white films, although green or yellow filters are commonly used to increase contrast.

Film speed is inversely related to grain size. Slow speed films (ASA 64 vs. 160) provide smaller grains and therefore images of higher resolution (can be enlarged and still maintain quality, such as projected images in a large auditorium). These are used for subjects that do not move, such as photographing images with a microscope or dead mice undergoing necropsy. Higher-speed films are useful for live mice that are walking around or for low-light work, such as immunofluorescence microscopy.

## IV. COPY STANDS AND CAMERA SUPPORTS

A hand-held camera with a fast film is adequate for photographing live mice moving around, especially when an electronic flash is used (see below). However, a support for the camera is useful so that the mouse can be allowed to run around, and yet you can follow it easily without fear of dropping the camera. A tripod is often adequate for this purpose. A more stable platform is useful for necropsy specimens, particularly when close-up work is required. A variety of commercially available copy stands can be used (Figure 6.1). These often have two to four arms with light sockets for tungsten photo flood lamps on either side of the table. Some of these lamp fixtures are fixed, while other types are adjustable. Moving the lights can reduce or eliminate shadows. Adjustment of the lights should be made, if possible, to optimize the image. Polarizing filters can be placed in front of the lights to act as a partial heat shield and on the camera's lens to reduce highlights.

Most camera stands provide a fixed arm that slides up and down on a support post. The arm has a base with a screw to attach the camera. The end of the arm can be fitted with a custom unit that has adjustable and interchangeable brackets to support two cameras. Our prototype was fabricated from two hasps used to lock doors. The slotted area that the U-bolt slipped through for the attachment of a padlock served as an adjustable port for the screw that supported the camera. Two such hasps

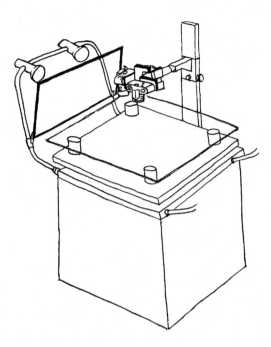

**FIGURE 6.1**  Copy stand with flood lamps and polarizing filter in place on left and a swing arm device to hold two 35 mm SLR cameras. Plate glass is raised above the level of the stage by four Plexiglas dowels. Black velvet can be placed on the stage under the glass to provide a shadowless background. (Drawn by Ingrid K. Sundberg.)

were welded together and attached to a piece of square, hollow steel that inserted into the arm of the copy stand. This was later replaced by a more refined unit (Figures 6.1 and 6.2). Such an arrangement permits the use of two cameras at the same time, one with black-and-white film and the other with color slide film. Once fit to the cameras, they can be interchanged without moving the specimen to provide rapid photography for both lecture presentation of slides and high-quality black-and-white negatives for generation of prints for use in publications. Even though the camera is mounted on the stand, a cable release is still needed to prevent camera shake, and blurring of the picture.

The area around the camera stand may be draped in inexpensive black cloth purchased locally from a fabric store. This approach may be necessary to eliminate daylight from a nearby window or fluorescent light from overhead lamps which will cause color shifts in the overall image generated, if the photographic area is not isolated.

Record keeping is the key to success. Although mice come in many different colors, it is easy to forget which ones went with which study. Therefore, keeping a record book next to the photographic set-up is important. Color slide film can be purchased in large quantities, along with mailers for development. The mailers come with an identification number, which should be recorded at the beginning of each

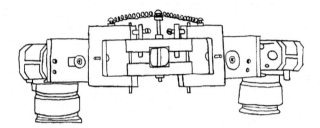

**FIGURE 6.2**    Detail of swing arm device in Figure 6.1. This permits two cameras with different film to be used interchangeably without moving the specimen. (Drawn by Ingrid K. Sundberg.)

list of photographs taken. Data collection can be limited to case number (see Chapter 4, Medical Record Keeping for Project Analysis), but it is best if information on the strain, age, sex, and type of lesion is provided. This will minimize the time needed to label slides once they are developed. A similar system can be used for black-and-white negatives. An alternative system, if there is a computer adjacent to the photography set-up, is to use a spreadsheet, such as Excel (Microsoft Corp., Redman, WA). The information listed above, plus a running number (becomes the negative record number), can be recorded. This number can be put on the slide or negative so that information can be retrieved rapidly. If the negative number is put on the back of prints, images can be managed years later when details are no longer easily obtained.

## V. BACKGROUNDS

The color of backgrounds affects color rendition of slides and apparent contrast in black-and-white images. Black backgrounds with black mice or white backgrounds with white mice make the mice difficult to see because they merge with the background. Grey or light blue backgrounds appear to be grey in black-and-white prints which is often perceived by reviewers as poorly printed images due to poor contrast. Light blue works well for dark mice in color films. Black works very well for mice that are not black.

If the plane of the specimen is raised above the level of the background, by use of a plate of 1/4-inch glass (with ground edges to avoid cuts) supported on Plexiglas cylinders 1 to 2 inches in diameter and 3 inches high, shadows in the background can be reduced or eliminated (Figure 6.1). If a large piece of black velvet (one square yard works well) is placed below the glass, all shadows are eliminated. This yields a very pleasing image for a light colored mouse on a solid black background.

Ordinary glass cleaner used sparingly and wiped away with paper towels is used before each series of photographs and frequently during the shoot to reduce dust that will settle on the glass. Once the subject is arranged on the glass, a small wad of moistened paper towel can be used around the edges of the specimen for last minute touch-ups. Cans of compressed air or inert gases, used to blow dust out of the field of view, are useful but not necessary. If used, point the tip away from the subject and toward the dust. A prolonged blast of gas with the nozzle near the glass creates a temporary pool of fluid that leaves a visible residue when the fluid dries. Powdered, disposable gloves are a big source of dust. Wash hands after putting on gloves to reduce this problem.

The glass plate should be larger than the specimen to allow it to be used several times by moving the subject to a clean part of the glass and moving the glass back under the camera. Since the glass is not fixed in place, it is easier to arrange the tissue or mouse on a clean area before centering it under the camera.

Cameras can reflect on the glass backgrounds and produce strange and distracting images in the field. Chrome-bodied cameras should be avoided, and black-bodied cameras should be used to minimize this. Regardless, chromed attachments, such as loops to hold a neck strap, may cause very minor, but annoying, reflections. These can all be reduced or eliminated by fabricating a mask out of black, nonglare paper or cardboard that fits over the lens.

## VI. PHOTOGRAPHING LIVE MICE

Photographing free-ranging, live mice requires patience. It is best to work out the distances for depth of focus, background material (color and texture), flash intensity, and film type well in advance by practicing with a dead mouse or simply a wad of paper. Work through several f-stops, distances, etc. Record every detail and condition of the test photographs while they are being taken. This part is tedious but necessary until the capabilities of the camera system are well understood. For slides, compare them side by side on a light-box, and the differences will be easier to see than if viewed singly through a slide viewer or projected on a screen.

Healthy, live mice are very active and can present many challenges for the photographer. A pleasing and unobtrusive background is the first issue of importance. The objective is to avoid a very light or dark shadow or blurry area in the background that will distract the viewer's attention from the image. This can be obtained by using a curved, solid, background, such as a large sheet of thin art board laid against a wall and curved to the front of the copy stand table. Colored paper in rolls can be hung from a wall and pulled down to a table as needed. This provides a renewable background to deal with urination and defecation, which create distracting shadows.

Mice will also run, jump, and try hard to escape. Physical barriers, such as use of an aquarium, limits their ability to escape and provides a container into which to look down on them. A variety of elevated pedestal arrangements often work well. Inverted beakers, cups, pipes, etc., forces the mouse to remain in a limited area. The glass beaker is easily cleaned if the mouse urinates or defecates. Live animal photographs are aesthetically pleasing, which has increasingly become a concern for publication of gross photographs.

Photographs of live mice are taken with a lens of longer focal length than that used with the photography stand (100 mm or larger compared with 50 to 100 mm for the copy stand). The longer focal length permits the photographer to remain farther from the mouse. A 100-mm macro lens is the minimum. It provides a greater working distance and therefore a little space for the mouse to walk or change posture without ruining the setting. With the lens focused at two feet, the mouse will fill most of the frame. If the mouse is particularly active, its tail can be held out of the scene to partially immobilize it.

The photoflood lamps on the normal photography stand are too hot for the live mouse. An electronic flash unit is perfect for a moving mouse. A small top-mounted flash unit will provide enough light in most cases for a 100-mm lens set at two feet. The farther the flash is set from the subject, the more light is needed, in which case the flash unit needs to be set at a higher intensity. A ringflash, which is a flash-tube fitted around the circumference of the lens, can be used, but it may make subjects look flat and relatively lifeless by eliminating most shadows.

Color daylight slide film is available in high speeds (ASA 100 and greater, KODAK Elite II; Eastman Kodak, Rochester, NY) and works very well with an electronic flash.

A device has been developed to permit live mice to be photographed repeatedly to track dorsal skin lesions. Mice are placed in essentially a three-sided box with a concave end into which they place their heads to avoid the intense light of the photofloods. The width can be adjusted to limit lateral mobility. Mice will remain essentially motionless.[1] If the mouse is tattooed to provide permanent reference points, skin lesions, such as induced papillomas, can be followed on a weekly basis to observe and photograph progression of disease.[2]

## VII. PHOTOGRAPHING NECROPSY SPECIMENS

Live mice move around, so many frames will have to be taken in the hope of obtaining an aesthetically pleasing, focused image that illustrates the lesions in question. However, multiple views are often useful because, until the image is selected, it can be difficult to determine which view is most useful. Also, if multiple views are taken, they can be used in subsequent papers or book chapters without copyright infringement concerns.

The specimen should be carefully prepared. Excess blood should be removed. Hair and other materials not normally found in the area being photographed should be removed. Organs should be displayed in an anatomically correct manner. This means not only as they are found in the animal but also in an orientation that is commonly presented in textbooks to minimize the viewer's need to determine the anatomy. A normal mouse may need to be placed adjacent to the mutant mouse. For example, mice with the mutation called situs inversus (gene symbol *iv*), have normal organs, but they are located on the side opposite normal. Simply turning over a slide would produce a correct image of a normal mouse. Displaying the viscera of a mutant mouse next to a normal mouse clearly illustrates the abnormality.[3]

Organs can be removed and photographed. Again, it may be important to have a normal organ adjacent to the abnormal organ to accentuate the abnormalities. The same criteria for presentation apply here as for the entire mouse. Clean specimens, clean glass table, clean background, attention to shadows and highlights are all important. Extension tubes can increase magnification for hard-to-see details. Moving lights around or adding light from a moveable floodlight can create or remove highlights and shadows that can make an abnormality show up or disappear. The apparently smooth surface of the liver can be seen to be pitted and irregular if the lighting is changed to highlight these features by creation of shadows or glare.

Dissection microscopes may be needed to view and photograph very small anatomic structures, such as defects in eyes or hair types on the skin. General approaches described under Photomicroscopy can be used. High-resolution images of small structures are often best evaluated by scanning electron microscopy. Specimen collection and preparation are described in Chapter 9, Ultrastructural Evaluation of Mouse Mutations.

A camera with manual focus can be used to take crisp images of very small features. These areas can later be cropped to eliminate areas that are not in focus. Depth of field is achieved by stopping down the diaphragm of a camera to reduce light coming in but increasing the area in front and behind a specimen that will be in focus. We set the cameras on f16 or f22 and adjust shutter speed to compensate for variations in light intensity. Since necropsy specimens do not move, long exposure times are possible.

## VIII. PHOTOGRAPHING DOCUMENTS AND RADIOGRAPHS

In addition to photographing dead mice, a copy stand can be used for photographing flat documents for seminars such as texts, electron micrographs, radiographs, and other such documents. The supporting glass should be cleaned carefully and the document laid down on the glass under the camera. A second piece of clean glass should be placed over the document to uniformly flatten it before the photograph is taken. To ensure the camera back is parallel to the glass plates, a small level can be placed on the camera back and its attachment adjusted. If a journal article is being photographed with print on both sides of the page, a solid black piece of paper can be placed behind the page being photographed to remove the bleed-through effect of print from the back side.

Black-and-white prints of electron micrographs can be taken this way using tungsten slide film with photo floodlamps to obtain slides for lectures. Radiographs, although black-and-white, are viewed with a light-box that most often contains fluorescent bulbs. The light-box can be placed on the copy stand, radiographs placed upon it and photographed using color slide film. The light source will cause a color shift, sometimes toward blue. Color correction filters can be used to obtain a true black-and-white slide, or black-and-white positive films can be used.

## IX. PHOTOMICROSCOPY

Taking pictures through a microscope well is an art, but the same basic approaches are used. Many books detail methods,[2,3] however, only simple approaches are needed to obtain high-quality results. One needs to become familiar with the microscope and camera and customize film, filters, lighting, etc., to obtain optimal results. Clean fields, evenly illuminated light grey backgrounds, and specimens free of artifacts are all that are needed. A small piece of dust not cleaned from a slide will create a poorly defined fuzzy image in the photograph that will ruin it completely. These types of images do get published, but editors and reviewers should not permit it!

As mentioned above, slow speed, fine-grain films are ideal for routine light microscopy. Tungsten type color films work well with tungsten–halogen light sources. The film will list the color temperature of the light source needed in the package insert. The microscope manual will provide instructions on how high to turn up the rheostat on the microscope to match this temperature. Most microscopes meet this criteria when the rheostat is turned to the highest level of illumination. Tungsten filaments, which are usually only found on very old microscopes, burn out quickly at this intensity. Halogen bulbs last for a long time but generate a lot of heat. Neutral density filters should be added to reduce the light's intensity to make using the microscope comfortable. Neutral density filters will not change the color of the light source.

Many color films (not Kodachrome, Eastman Kodak) can be optimized by a special filter called a didymium multiband glass filter.[4] Black-and-white films will yield marked improvement in contrast with hematoxylin and eosin stained sections when a green filter (Wratten #11, Eastman Kodak; IF550, Olympus Corp. America, Precision Instrument Div., New Hyde Park, NY) is added. Polarizing filters can be added above and below the specimen for demonstrating birefringence and recorded on either black-and-white or color films.

Köhler illumination should be set for each objective. Most research-level microscopes can be adjusted by closing down the field diaphragm, focusing the specimen, and then focusing the condenser on the leaves in the field diaphragm. The condenser is then centered. The field diaphragm is opened to the edge of the visible field, and the aperture diaphragm set for the optimal image. Then the photograph is taken.

Fields should be selected based on quality of section, whether they illustrate the abnormality of interest, and are clear of sectioning artifacts, dirt, etc. Ideally the section should be oriented in an anatomically correct position.

Research-quality microscopes usually have automated metering systems, so simply entering the film speed and reciprocity factor on older models (a table is included in the manual for reciprocity factors) or inserting the film, if it has metal strips on the cassette that will automatically be interpreted by the meter, will set up the photographic details. This usually works well for color film but is less reliable for black-and-white film. Black-and-white film can be purchased in bulk and a roll exposed to determine the optimal film speed (ASA or ISO rating).

Vibration is the greatest problem in photomicroscopy. The microscope should be placed on a strongly supported workbench or desk in a vibration-free room. Vibrations can be minimized by using marble tables on concrete floors or on various types of air tables (Technical Manufacturing Corp., Peabody, MA) that attempt to isolate the vibration from the microscope.

Darkfield photomicroscopy is useful for evaluating some anatomic structures in normal histologic sections, but it is particularly useful for autoradiographs of tissue sections. Approaches to this are discussed in Chapter 11, Radiolabeled cRNA and *In Situ* Hybridization.

## X. CONCLUSIONS

Medical microscopy is an art that can be mastered with perseverance and talent. The general points provided here serve as starting points. Combined with the reference lists for more detailed approaches,[4-7] investigators should be able to set up a system to generate high-quality images with very little time and effort.

## ACKNOWLEDGMENTS

This work was supported by a grant from the National Cancer Institute (CA34196). The swing arm device for holding two cameras at once was designed and built by Bud Sorensen of Bar Harbor, Maine, for our laboratory.

## REFERENCES

1. Binder, R.L., Nonstressful restraint device for longitudinal evaluation and photography of mouse skin lesions during tumorigenesis studies, *Lab. Anim. Sci.*, 46, 350, 1996.
2. Binder, R.L., Gallagher, P.M., Johnson, G.R., Stockman, S.L., Smith, B.J., Sundberg, J.P., and Conti, C.J., Evidence that initiated keratinocytes clonally expand into multiple existing hair follicles during papilloma histogenesis in SENCAR mouse skin, *Mol. Carcinogenesis,* 20, 151, 1997.
3. Sundberg, J.P. and Collins, R., Animal models of human disease: situs inversus mouse, a model of Kartagener's syndrome, *Comp. Pathol. Bull.*, 24, 2, 1992.
4. Delly, J.G., *Photography Through the Microscope*, 9th edition, Eastman Kodak Company, Rochester, NY, 1988.
5. Olympus Optical Co., Ltd. *How to Improve Photography Through the Microscope,* Tokyo, 1986.
6. McGavin, M.D. and Thompson, S.W., *Specimen Dissection and Photography,* Charles C. Thomas, Springfield, IL, 1988.
7. LeBeau, L.J., Pederson, E.K., Parshall, R.F., and Beluhan, F.Z., Photography of small laboratory animals. Part two: techniques and procedures, *J. Biol. Photogr.*, 55, 99, 1987.

# 7 Comparative Pathology and Animal Model Development

*John P. Sundberg and Lloyd E. King, Jr.*

## CONTENTS

## I. INTRODUCTION

Once a stable colony of either spontaneous or induced mutant mice is established, the challenges of characterizing the phenotype and comparing it to similar phenotypes in well-documented diseases of humans and domestic animals begins. Collaboration between veterinarians and physicians can provide a unique resource to both expedite this process and provide insight and accuracy to the project. The veterinarian, in particular a veterinary pathologist trained in the normal and abnormal anatomy and physiology of the mouse, understands the mouse and has a background in general mammalian pathophysiology. This not only provides expertise with which to evaluate the mutation but also comparative information from domestic mammals that gives the study a phylogenetic perspective. Physicians work with humans on a daily basis and are familiar with their pathophysiology, which is similar but also uniquely different from that of mice. Working together, putative lesions apparent to the physician may be determined to be normal by the veterinarian, thereby avoiding a potentially serious error. For example, during initial evaluation of the mouse model for alopecia areata,[1] we found that actively growing hairs were surrounded and infiltrated by inflammatory cells. Dr. King, the physician, immediately recognized the similarity of this lesion to the human disease, alopecia areata.[2] He interpreted the small hair follicles adjacent to those affected as being reduced in size as a sequela to the disease process, a common feature for a number of human hair diseases.[2] Dr.

Sundberg, the veterinary pathologist, knew that the small hair follicles were normal telogen, the resting phase of the hair cycle for mice. Although he recognized the inflammatory process in the anagen follicles, he was not aware of the specific disease since at the time very little was known of alopecia areata-like syndromes in domestic animals. Working together, Drs. King and Sundberg rapidly and correctly interpreted the lesions, and this has become a standard model for evaluating alopecia areata.

Many spontaneous mouse mutations have been identified and characterized for decades, including cloning and characterizing the mutated gene. Some of these, like hairless (gene symbol: *hr*) and nude (gene symbol: *Hfh11^nu*) have recently been used as tools to identify homologous human mutations by using the mouse gene sequence to evaluate DNA from human patients with similar phenotypes.[3-6] Interdisciplinary collaborations were necessary to identify the similarities between the species, find the patients, and investigate the molecular biology. This type of approach is highly productive and is setting the standard for phylogenetic investigations.

## II. CRITERIA FOR ANIMAL MODELS

The value of a mutant mouse historically depended upon whether or not it could be developed as a model. This has changed dramatically with the advent of transgenesis and targeted mutagenesis where specific genes can be totally or partially inactivated or overexpressed generally or in specific organs. The mouse is now used more as a "living test tube" to establish the real function of a gene *in vivo*. Regardless, a number of historical criteria exist for establishing the value of animal models (Table 7.1).[7-9] The most common criteria used is whether or not the animal disease resembles that of a particular human disease. Although this is desirable, this is not the only value of such models. Models are useful if they can be used to answer very specific questions regardless of whether or not they have correlation to human diseases. For example, the beige mouse mutation in the lysosomal trafficking regulator gene (gene symbol: *Lyst^bg*) produces a phenotype that is essentially identical to the human disease known as Chediak–Higashi syndrome.[10] The phenotype in humans and mice is mimicked in many other species so this genetic defect is phylogenetically highly conserved. In spite of this, Chediak-Higashi syndrome is so rare in humans and domestic animals that beige mice are rarely, if ever, used as animal models for the disease. Rather, since they have very specific immunological and readily identifiable coat-color defects, they are used in numerous basic science studies.[10]

The simplistic approaches of studying a single gene mutation on one inbred genetic background are being circumvented by very complicated and sophisticated manipulations when multiple mutations are combined on the same inbred background to do standardized studies in which factors or cells are added or subtracted to answer very specific mechanistic questions. For example, historically, to investigate the role of hormones on hair growth one would study hair follicles in patients with various endocrine diseases. Today it is possible to graft human skin with hair follicles onto the back of various immunodeficient mice that will accept such xenografts. Such mice can be created to carry mutant genes that create deficiencies in various hormones, such as mutant mice that lack thyroid hormones (hypothyroid, thyroid-stimulating hormone receptor gene, gene symbol: *Tshr^hyt*), androgens and

---

**Table 7.1**
**Commonly Used Criteria for Animal Model Systems[6-8]**

Accurately reproduce the disease (usually means the human disease)

Available to researchers (mice are one of the easiest species to use in research)

Exportable to international researchers (international repositories make mice ideal for this)

Species that have multiple offspring for genetic-based diseases

Sample size adequate (microchemistry has almost eliminated this as a criterion)

Animal facilities generally available

Ease in management and safe handling

Disease is phylogenetically conserved (models are available in multiple species)

Survivability (molecular methodology permits *in utero* studies of lethal mutations so this is no
  longer a major criterion)

---

estrogens (hypogonadal, the mutation in the gonadotropin-releasing hormone gene, gene symbol: *Gnrh^hyp*), androgen receptors (the testicular feminization mutation, in the androgen receptor gene, gene symbol: *Ar^Tfm*), and many others.[10] Hormones can easily be added back to replace those missing.[11]

A seemingly endless array of experimental situations can be set up to exploit inbred mice with various mutations. However, the fundamental question arises when a new spontaneous mutation occurs in a colony or a new transgenic or targeted mutation is created. How does one compare this new mutation with the myriad of human diseases that have been defined over the centuries?

## III. INFORMATION RESOURCES

As stated above, a knowledge of normal anatomy, and spontaneous, induced, and infectious diseases of mice is needed to both interpret the phenotype under investigation and to compare it with specific human diseases. This is beyond the scope of this book and also that of most other textbooks. The key to dealing with these topics is to have access to major reference works that are relatively recent and contain the details needed. To that end, a list of references is provided.[12-32] Furthermore, much information is becoming available on the world wide web. The Mouse Genome Database (MGD; hppt://www.informatics.jax.org) is maintained at The Jackson Laboratory and provides a large amount of information on the genotype and phenotype of mice, with connections to Online Mendelian Inheritance of Man, a similar database on human genetic diseases and other databases as well. The MGD is being expanded and updated regularly. To use this database efficiently in a search, a list of major phenotypes is available (Table 7.2). A pathology home page has been added, and a new resource, Mouse Tumor Biology Database, will have information on the pathology, incidence, biology, and genetics of mouse tumors including photomicrographs of the lesions. All of this can be accessed on the web (http://www.jax.org). A number of pathologists have been creating web sites (http://www.ncifcrf.gov/vet-path) focused on mouse pathology, and these will become an ever-increasing resource.

**Table 7.2**

**Phenotypic Criteria Available on the Mouse Genome Database (Genes, Markers, and Phenotypes Query Form) for Searching for Similar Mouse Mutations.[10]**

| | |
|---|---|
| Anatomical: | Eye |
| | Internal defects of viscera |
| | Skeleton |
| | Skin and hair texture |
| | Tail and other appendages |
| Biochemical: | Co-enzymes |
| | Enzymes |
| | Homeobox |
| | Ligands |
| | Other Proteins |
| | Receptors |
| | Transcription factor |
| Immunological: | Cell surface antigens |
| | Immune defects |
| | Other immunological |
| Markers: | DNA sequences, transgenes |
| | Pseudogene |
| | Related sequence |
| | RNA |
| | Transgenes |
| Neurological: | Inner ear and circling behavior |
| | Neurological and neuromuscular |
| | Other behavioral |
| Other: | Miscellaneous |
| | Unnamed loci |
| Physiological: | Color and white spotting |
| | Endocrine defects, hormones, growth, obesity |
| | Hearing |
| | Hematological |
| | Reproductive organs, sterility |
| Vital: | Endogenous viruses |
| | Homozygous lethality or sublethality |
| | Oncogenes and viral integration sites |
| | Viral, disease, and tumor resistance |

## IV. COMPARISON TO HUMAN DISEASES

Clinicians and preclinical scientists in pharmaceutical companies often have one criterion that needs to be met before a mutation can be considered to be a useful model. This criterion is simply whether the mutant mouse will respond to commonly used treatments for the human disease. This criterion in itself is not a true test for the value of a model because drug metabolism can vary dramatically between species. A very valuable mouse model may be discarded on the basis of only one type of treatment if the compound cannot be metabolized to an active form in mice because they lack an enzyme unrelated to the mutation. Other treatments may work perfectly well. Alternatively, it is also possible that a minor subgroup of human patients also lacks the enzyme and will not respond as anticipated.

An efficient way to determine if a new mutation is a likely candidate for a new model for a specific human disease would be to review the preliminary findings with a physician to narrow the differential diagnosis. Ideally, this review should be done collaboratively with a specialist. A thorough literature search, beginning with such databases as the Mouse Genome Database or Online Mendelian Inheritance of Man (see above), will often yield many helpful references on the disease. This literature review will provide data to create a table of all features reported to occur with the human disease and rank them in terms of importance or frequency. From these data, a matching list for the mouse mutation can be developed. This matching list provides a simple and rapid way to compare features and provides criteria to search for in the mutant mouse mutation and to determine which clinical features or clinical chemistry criteria should be investigated. For example, a mouse with hair loss or alopecia with specific hair fiber abnormalities can be compared with an apparently homologous human disease with the same types of hair fiber defects. This approach was used to determine that the lanceolate hair (gene symbol: *lah*) mouse mutation resembled the human disease called Netherton's syndrome.[33] Evaluation of the literature revealed that many human Netherton's syndrome patients have elevated levels of IgE. Evaluation of IgE levels in lanceolate hair and its newly discovered allele, lanceolate hair-J, revealed that both mutations were characterized by elevations of IgE and that one was far more elevated that the other.[34] These data provide evidence of a dichotomy between alleles that affect IgE levels that may help us understand variations in IgE levels and clinical features found in the human syndrome. Conversely, criteria found in the mouse mutants but not reported in obscure human homologs of the disease may point the clinician to abnormalities in the human that were previously overlooked. For example, the mouse mutation known as hairless (*hr*) has been studied for nearly 75 years and is known to have immunologic abnormalities. Identification and sequencing of the mutated mouse gene led to molecular studies in the human equivalent, papular atrichia.[3,4,6,35] Immunologic abnormalities in human patients are only now being investigated based on knowledge gained from studying the mouse.

For these reasons, it is important to do a thorough and unbiased workup of the mouse mutant under investigation, using the collective knowledge already in hand for an apparently homologous disease, and also not be too quickly dissuaded by novel findings.

## V. INBRED VS. OUTBRED

Inbred animals, by definition, are the end result of 20 or more controlled brother × sister matings, resulting in near genetic identity between all members of the strain (See Chapter 13, Repositories of Mouse Mutations and Inbred, Congenic, and Recombinant Inbred Strains). Because inbred mice can be raised in a controlled environment with a defined pathogen status (specific pathogen free colony), most of the mice in that colony will maintain a remarkably similar phenotype.[8,36,37] There may be sex-linked mutations, or sex may influence expression of a mutation, but results are highly predictable once identified. If a mutation arises for whatever reason in such a colony, its effect usually will be remarkably homogeneous in that inbred colony. If the mutant mice are crossed with clinically normal mice from an unrelated strain, the phenotype can disappear, get worse, change completely, or change in very subtle ways. These changes can be stabilized by continual backcrosses onto the other strain until 10 or more generations of backcrossing have occurred, at which time a congenic strain has been created. This is how mutant genes are transferred from one strain to another. Thanks to the foresight of Dr. Elizabeth Russell, many mutations have been transferred onto the C57BL/6J inbred strain. Different mutations on these congenic strains can be combined by breeding to determine the effect of interactions between the mutations without concern for other modifier genes that would be injected if unrelated strains were used. This congenic strain approach allows very sophisticated experiments to be designed.

Humans and domestic animals tend to be outbred or partially inbred, and they live in microbiologically "dirty" environments (exposed to a variety of different pathogenic organisms). Many genes are intermixed from both parents to produce offspring in a population that has little genetic homogeneity. Few genes function totally independently of all others. Those genes with some effect on other genes are called *modifier genes*. If a mutation occurs and creates a disease, the modifiers will control how severe the disease is or what specific lesions are found in any one individual. The result can be that parents with minor and insignificant forms of a disease may have a child that is barely able to stay alive. An example of these phenomena are seen in targeted mutations of epidermal growth factor receptor, in which severity ranging from hair fiber abnormalities to lethality is dependent upon the genetic background of the mouse.[38,39] These features of genetic diseases in an outbred population make it extremely difficult to do direct comparisons between inbred and outbred species to develop animal models and convince clinicians of their validity. Rather, the inbred mouse strains carrying mutations probably represent specific subtypes of any human genetic disease. Because of the high degree of genomic homology between mice and humans, it continues to be easier and faster to map, clone, and sequence mouse genes and then find the human homolog, rather than the inverse. Identifying modifier genes is almost impossible with humans but has become a standard part of mouse genetics. These are powerful tools to help us understand the complexities of the genetic basis of disease, and this is redirecting our fundamental understanding of medicine.

## VI. CONCLUSIONS

Evaluation of the genetics and phenotype of spontaneous and induced mouse mutations is no longer a simple operation that can be done by a single individual. Groups of specialists working together are needed to carefully define the phenotype in the mouse, compare it with similar diseases in humans, integrate species differences, and correlate these findings with genetic and biochemical findings. In this way the spontaneous and induced mouse mutations will be carefully and rapidly evaluated and appropriately compared with human diseases to produce useful models that can be applied to answer specific problems.

## ACKNOWLEDGMENTS

This work was supported by grants from the National Institutes of Health (CA34196, AR43801, and RR8911).

## REFERENCES

1. Sundberg, J.P., Cordy, W.R., and King, L.E., Jr., Alopecia areata in aging C3H/HeJ mice, *J. Invest. Dermatol.*, 102, 847, 1994.
2. Olsen, E.A., *Disorders of Hair Growth*, McGraw-Hill, Inc., Health Professions Division, New York, 1994.
3. Ahmad, W., ul Haque, M.F., Brancolini, V., Tsou, H.C., Lam, H., Alta, V.M., Owen, J., deBlaquiere, M., Frank, J., Cserhalmi-Friedman, P.B., Leask, A., McGrath, J.A., Peacocke, M., Ahmad, M., Ott, J., and Christiano, A.M., Alopecia universalis associated with a mutation in the human hairless gene, *Science*, 279, 720, 1998.
4. Panteleyev, A.A., Paus, R., Ahmad, W., Sundberg, J.P., and Christiano, A.M., Molecular and functional aspects of the hairless (*hr*) gene in laboratory rodents and humans, *Exp. Dermatol.*, 7, 249, 1998.
5. Jorge, F., Pignata, C., Panteleyev, A.A, Prowse, D.M, Baden, H., Weiner, L., Gaetaniello, L., Ahmad, W., Pozzi, N., Cserhalmi-Friedman, P.B., Gordon, D., Ott, J., Brissette, J., and Christiano, A.M., Congenital alopecia and severe T-cell immunodeficiency associated with a mutation in the human nude gene, *Science*, submitted.
6. Sundberg, J.P., The hairless (*hr*) and rhino (*hr^{rh}*) mutations, chromosome 14, in Sundberg, J.P. (Ed.), *Handbook of Mouse Mutations with Skin and Hair Abnormalities. Animal Models and Biomedical Tools*, CRC Press, Inc., Boca Raton, 1994, 291.
7. Leader, R. W. and Padgett, G. A., The genesis and validation of animal models, *Am. J. Pathol.*, 101, S11, 1980.
8. Sundberg, J.P., Inbred laboratory mice as animal models and biomedical tools: general concepts, in Sundberg, J.P., Ed., *Handbook of Mouse Mutations with Skin and Hair Abnormalities. Animal Models and Biomedical Tools*, CRC Press, Inc., Boca Raton, 1994, 9.
9. Sundberg, J.P., Animal models for papillomavirus research, in *Viruses in Human Tumors*, Hofschneider, P.H. and Munk, K., Eds., Karger, Basel, 1987, 11.
10. Mouse Genome Database. The Jackson Laboratory, Bar Harbor, ME, http://www.informatics.jax.org, September 1998.
11. Sundberg, J.P. and King, L.E. Jr., Mouse mutations with skin and hair abnormalities as animal models for dermatological research, *J. Invest. Dermatol*, 106, 368, 1996.
12. Feldman, D.B. and Seely, J.C., *Necropsy Guide: Rodents and the Rabbit*, CRC Press, Inc., Boca Raton, 1988.
13. Gude, W.D., Cosgrove, G.E., and Hirsch, G.P., *Histologic Atlas of the Laboratory Mouse*, Plenum Press, New York, 1982.
14. Loeb, W.F. and Quimby, F.W., *The Clinical Chemistry of Laboratory Animals*, Pergamon Press, New York, 1989.

15. Popesko, P., Rajtova, V., Horak, J., *A Color Atlas of Anatomy of Small Laboratory Animals, Volume 2: Rat, Mouse, Hamster,* Wolfe Publishing Ltd., London, 1990.

16. Darai, G., *Virus Diseases in Laboratory and Captive Animals,* Martin Nijhoff Publishing, Boston, 1988.

17. Frith, C.H. and Ward, J.M., *Color Atlas of Neoplastic and Non-neoplastic Lesions in Aging Mice,* Elsevier, Amsterdam, 1988.

18. Jones, T.C., Mohr, U., and Hunt, R.D., *Monographs on Pathology of Laboratory Animals. Endocrine System,* Springer-Verlag, Berlin, 1983.

19. Jones, T.C., Mohr, U., and Hunt, R.D., *Monographs on Pathology of Laboratory Animals. Digestive System,* Springer-Verlag, Berlin, 1985.

20. Jones, T.C., Mohr, U., and Hunt, R.D., *Monographs on Pathology of Laboratory Animals. Respiratory System,* Springer-Verlag, Berlin, 1985.

21. Jones, T.C., Mohr, U., and Hunt, R.D., *Monographs on Pathology of Laboratory Animals. Urinary System,* Springer-Verlag, Berlin, 1986.

22. Jones, T.C., Mohr, U., and Hunt, R.D., *Monographs on Pathology of Laboratory Animals. Genital System,* Springer-Verlag, Berlin, 1987.

23. Jones, T.C., Mohr, U., and Hunt, R.D., *Monographs on Pathology of Laboratory Animals. Integument and Mammary Glands,* Springer-Verlag, Berlin, 1989.

24. Jones, T.C., Ward, J.M., Mohr, U., and Hunt, R.D., *Monographs on Pathology of Laboratory Animals. Hematopoietic System,* Springer-Verlag, Berlin, 1990.

25. Jones, T.C., Mohr, U., and Hunt, R.D., *Monographs on Pathology of Laboratory Animals. Cardiovascular and Musculoskeletal System,* Springer-Verlag, Berlin, 1991.

26. Jones, T.C., Mohr, U., and Hunt, R.D., *Monographs on Pathology of Laboratory Animals. Eye and Ear,* Springer-Verlag, Berlin, 1991.

27. Jones, T.C., Mohr, U., and Hunt, R.D., *Monographs on Pathology of Laboratory Animals. Nervous System,* Springer-Verlag, Berlin, 1991.

28. Lindsey, J.R., Boorman, G.A., Collins, M.J., Van Hoosier, G.L., and Wagner, J.E., *Infectious Diseases of Mice and Rat,* National Academy Press, Washington, D.C., 1991.

29. Mohr, U., Dungworth, D.L., Capen, C.C., Carlton, W., Sundberg, J., Ward, J.M., *Pathobiology of the Aging Mouse, Volumes 1 and 2,* ILSI Press, Washington, D.C., 1996.

30. Percy, D.H. and Barthold, S.W., *Pathology of Laboratory Rodents and Rabbits,* Iowa State University Press, Ames, 1993.

31. Sundberg, J.P., *Handbook of Mouse Mutations with Skin and Hair Abnormalities: Animal Models and Biomedical Tools,* CRC Press, Inc., Boca Raton, 1994.

32. Turusov, V.S. and Mohr, U., *Pathology of Tumours in Laboratory Animals, Vol. 2. Tumours of the Mouse,* 2nd ed., International Agency for Research on Cancer, Lyon, France, 1994.

33. Montagutelli, X., Hogan, M.E., Aubin, G., Lalouette, A., Guenet, J.-L., King, L.E., and Sundberg, J.P., Lanceolate hair (*lah*): a recessive mouse mutation with alopecia and abnormal hair, *J. Invest. Dermatol.,* 107, 20, 1996.

34. Sundberg, J.P., Boggess, D., Bascom, C., Limber, B.J., Shultz, L.D., Sundberg, B.A., King, L.E., Jr., and Montagutelli, X., Lanceolate hair-J (*lah^J*): A spontaneous mouse mutation resembling Netherton's syndrome, *Am. J. Pathol.,* submitted.

35. Sundberg, J.P., Dunstan, R.W., and Compton, J.G., Hairless mouse, HRS/J *hr/hr,* in *Monographs on Pathology of Laboratory Animals. Integumentary and Mammary Glands,* Jones, T.C., Mohr, U., and Hunt, R.D., Eds., Springer-Verlag, Heidelberg, 1989, 192.

36. Sundberg, J.P., Mouse mutations: Animal models and biomedical tools, *Lab Animal*, 20, 40, 1991.
37. Sundberg, J.P., Conceptual ideas on the use of mouse mutations as animal models, *Lab Animal*, 21, 48, 1992.
38. Threadgill, D.W., Dlugosz, A.A., Hansen, L.A., Tennenbaum, T., Lichti, U., Yee, D., LaMantia, C., Mourton, T., Herrup, K., Harris, R.C., Barnard, J.A., Yuspa, S.H., Coffey, R.J., and Magnuson, T., Targeted disruption of mouse EGF receptor: effect of genetic background on mutant phenotype, *Science,* 269, 230, 1995.
39. Sibilia, M. and Wagner, E.F., Strain-dependent epithelial defects in mice lacking the EGF receptor, *Science,* 269, 234, 1995.

# 8 Kinetics and Morphometrics

*Richard S. Smith, Gregory Martin, and Dawnalyn Boggess*

## CONTENTS

## I. KINETICS

Abnormalities in the size, shape, and architecture of organs at the gross and microscopic levels have been evaluated subjectively by pathologists for decades. Refining these observations in longitudinal characterization studies to aid in determining the pathogenesis or quantification for complex gene mapping studies has been simplified using computer-assisted techniques. This chapter will provide an overview of commonly used approaches for quantifying DNA synthesis rates, cell turnover, and approaches to measuring specimens.

Determination of relative rates of cell proliferation is often important in evaluating mouse mutations. For example, when phenotypic differences are subtle, it may become necessary to quantitate cellular activity in order to differentiate between homozygotes, heterozygotes, and wild-type mice. The ability to determine significant quantitative differences may be particularly important with phenotypes that are influenced by modifier genes (quantitative traits). These measurements may also be useful in deciding whether a thickened cell layer is due to enhanced proliferation or to a decrease in cell death.[1] Measurements may require simple counts of labeled cells or more complex morphometric analyses. The purpose of this chapter is to outline commonly utilized approaches for labeling cells and analyzing tissues.

DNA synthesis, mitotic indices, and cell cycle rates may be measured by a variety of methods. Some require pulse-labeling methods, while others can utilize

**111**

routinely collected, paraffin-embedded sections. These methods are applicable to all organ systems. Actively growing cells pass through the cell cycle, usually considered to consist of four steps: $G_1$, the relatively inactive period following mitosis; S, the stage of DNA synthesis; $G_2$, the postsynthetic/premitotic phase; and M, the mitotic phase. Pulse-labeling methods indicate the number of cells in the DNA synthetic phase (Figure 8.1).[2]

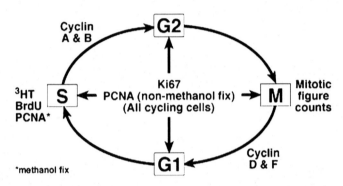

**FIGURE 8.1** Cell cycle.

## A. TECHNIQUES AND PROTOCOLS

1. *Radioactive labeling* is a technique that has been used successfully for the past 50 years. Thymidine is incorporated into DNA during the S-phase of the cell cycle and can be isotopically labeled with tritium ($^3$H) or carbon ($^{14}$C). Usually, a single dose is administered intraperitoneally, followed by tissue collection after 1 to 2 hours. Depending on experimental design, multiple pulses of $^3$H-thymidine may be required. Multiple injections, followed by a four-week chase period, are utilized to identify the location of conjunctival and cutaneous stem cells.[3,4] The usual dosage for mice is 1 to 5 µCi/gm body weight of $^3$H-thymidine.[1,3,5] Any cell in the S-phase exposed to the $^3$H-thymidine will take up the isotope. It should be emphasized that standard precautions and techniques for working with radioactive materials should be observed at all times.

Paraffin sections are prepared and covered with a liquid photographic emulsion under darkroom conditions (Kodak NTB-2, Eastman Kodak Co., Rochester, NY). The slides are usually exposed for 30 days at 4°C (to optimize visualization, extra slides can be cut and exposed for 25, 30, 35, and 40 days), developed (Kodak D19 developer, Eastman Kodak Co.), and counterstained with hematoxylin.[5] Cells with more than four grains/nucleus are considered positive and may be counted manually and expressed as counts per unit area (Figure 8.2A). The disadvantages of this technique include (1) the training required to work with radioactive materials, (2) potential biohazards, and (3) the wait of four to six weeks for exposure of the emulsion. In addition, there is always diffuse background tissue labeling that may

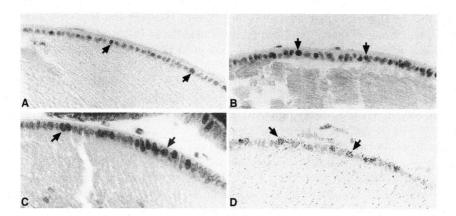

**FIGURE 8.2** A. Normal mouse lens epithelium stained with Ki-67 antibody. Only seven cells are stained. B. Normal mouse lens epithelium stained with PCNA; nearly all cells are positive. C. Normal mouse lens epithelium stained with BrdU; half of the cells are stained D. Normal mouse lens epithelium after tritiated thymidine exposure; nearly half the cells demonstrate uptake of the radiolabel. Arrows indicate selected positive cells. Original magnification × 630.

be confusing in determining positive vs. negative cell labeling, although the visualization is enhanced by the use of dark-field microscopy.

2. *Bromodeoxyuridine* (BrdU): BrdU offers a nonradioactive alternative for labeling cells entering the "S" phase. Mutant and control mice are injected intraperitoneally with 50 µg/g body weight BrdU. After one hour, mice older than 12 days are euthanized by $CO_2$ asphyxiation and tissues are removed and placed in the appropriate fixative for 12 to 24 hours (see Chapter 5). Newborn mice, up to seven days of age, are held for two hours before tissue collection. After routine histological processing, the blocked tissues are sectioned at 5 µm and stained with hematoxylin and eosin for routine histologic evaluation. Sequential unstained sections are placed on slides coated with poly-L-lysine to minimize folding and tissue loss and prepared for immunohistochemical staining to detect BrdU. The tissues are hydrated routinely, a primary rat monoclonal antibromodeoxyuridine antibody is applied (diluted 1:50; Accurate Antibodies, Sera Labs, Westbury, N.Y.) and processed, using a commercially available kit (Vector Laboratories, Burlingame, Calif.). The number of positive staining nuclei (dark brown nucleus) is counted (Figure 8.2B). Labeling of cells with either BrdU or ³H-thymidine produces comparable results.[6]

3. *Proliferating cell nuclear antigen* (PCNA) is an acidic nuclear protein that is synthesized prior to mitosis and usually designated as PCNA to distinguish it from other members of the cyclin superfamily.[7] No previous treatment of the experimental animal is required for demonstration of PCNA expression. Cells fixed in either 4% formaldehyde in $0.1M$ cacodylate buffer or in absolute methanol demonstrate the most pronounced staining (Figure 8.2C).[8] The method of fixation is important, since PCNA labeling with methanol fixation is limited to the "S" phase cells, while other

fixatives produce staining of any cycling cell and is, therefore less specific.[9] Questions have also been raised concerning its validity (compared with BrdU) for assessing tumor-related mortality.[10] Despite these objections, PCNA staining offers the unique advantage that archival paraffin-embedded tissues can be used to identify cycling cells.[11,12]

4. *Ki-67* is a monoclonal antibody that detects a nuclear antigen that is a nonhistone protein present in proliferating cells but absent in quiescent cells.[13] It has been used for assessing prognostic categories in some tumors and is believed to be a more specific indicator of cell proliferation than PCNA, even though the labeling is not restricted to "S" phase cells.[14] The number of labeled cells is said to correlate well with tritiated thymidine labeling.[9] In addition, our experience indicates that fewer nuclei are stained with Ki-67 than with any of the previously described techniques. As an example, (Figure 8.2A), the lens epithelium demonstrates prominent staining, but in the same specimen, proliferating cells of the epidermis failed to stain, even though positive results were obtained with tritiated thymidine, BrdU, and PCNA in sections from the same tissue.

Ki-67 is visualized using standard avidin–biotin–peroxidase techniques from commercially available kits. Slides are deparaffinized and rehydrated through xylenes and absolute alcohols to water. The sections are then treated with trypsin (Zymed Cat #00-3003) at 37°C in a moist chamber for 30 to 45 minutes and rinsed in several changes of PBS. The slides are next placed in 0.3% $H_2O_2$ in methanol for 20 minutes at room temperature to block endogenous peroxidase. The slides are washed in distilled water 3 times for 2 minutes each, then placed in a microwave-safe, solvent-resistant staining dish. Sodium citrate buffer ($0.01M$) is prepared, titrated to pH 6.0, and added to the staining dish to completely cover the slides. The slides are microwaved on high in 2-minute intervals for a total of 8 minutes, and checked regularly to ensure that slides remain immersed in the buffer. After heating, slides are cooled in buffer for 30 minutes, then washed in PBS 3 times at 3 minutes each. Using the Vectastain Rabbit ABC kit (Vector Laboratories, Cat #PK-4001), slides are treated, using the standard protocol provided by the manufacturer. Primary antibody is incubated overnight at 4°C. (Polyclonal Rabbit anti-Human Ki-67 Antigen. Zymed Cat. #ZS18-0191). The reaction product is developed with diaminobenzidine (DAB).

5. *Mitotic rates* can be measured and reported as the number of mitotic figures per high-power field or for a defined area. This provides an alternate or adjunct to the previously described methods, since it measures a different part of the cell cycle (mitotic phase). Although many histochemical stains are available, hematoxylin alone provides the best contrast for evaluating and counting mitotic figures (Sundberg, unpublished observations).

6. *General problems of interpretation*: In both tritiated thymidine and BrdU labeling, a comparison is usually made between tissues from normal controls vs. mutant mice. Care must be taken that the "controls" correctly fit the term. An example would be a gene that produces a heterozygote effect: an animal that is phenotypically normal by clinical examination may still have microscopic phenotypic differences from a true wild-type mouse. Genotyping, if possible, will control for this. Location of tissue sampled must be held constant between control and affected mice. A sample

of dorsal skin cannot be compared with a sample of ventral skin. In many organs, such as the eye, tissue orientation and plane of section also need to be constant. When tissue, such as the corneal epithelium is measured, the results will be different if the section used contains central or peripheral cornea.[1] An important part of the experimental protocol is to define *what* is being measured, *where* it is being measured, and *how* it is being measured. If these parameters are not strictly controlled, conclusions drawn will be meaningless.

## B. DATA COLLECTION AND ANALYSIS

Since many of the described labeling techniques demonstrate cells in the "S" phase of growth, these methods are often used to provide quantitative measurements of cell proliferation. The process of mitosis is also easily detected, enabling calculation of the mitotic index. In both instances, the first step is to define the area of measurement (e.g., labeled cells/linear mm of tissue; labeled cells touching the basement membrane; labeled cells per 400× field, etc.) Both labeled cells or cells in mitosis can be quantitated in this fashion. A similar approach would be to count the number of labeled or mitotic cells/1000 total cells counted.[15] It should be emphasized that well-defined parameters of measurement must be established and used with care, if useful quantitative data is the goal. A review of the literature for each will define the commonly used criteria for any particular anatomic structure, but is beyond the focus of this chapter. Some structures such as the hair follicle infundibulum, are rarely studied for kinetics. A variety of approaches can be used, and if well defined, they will provide useful and valid data.[16] Utilization of morphometric techniques may be useful in quantitation of cell proliferation rates.

## II. MORPHOMETRICS

The term morphometrics refers to a variety of techniques designed to quantitate measurable differences in cells and tissues. In some instances, simple techniques are capable of providing useful results that clearly demonstrate an effect. As more sophisticated computer software and hardware become available, increasingly complex measurements, such as three-dimensional analysis, antigen quantitation,[17-19] and estimation of total numbers of specific structures in a tissue (e.g., renal glomeruli) have become possible.[20]

## A. SIMPLE MORPHOMETRICS

In many experimental designs there is a need to measure the differences in tissue layer thickness between control and mutant mice to determine if the visually observed difference is statistically significant. In order to make such measurements meaningful, tissue boundaries must be well-defined. The skin is an example of tissue where such boundaries are easily seen. For example, full skin thickness may be defined as the distance between the epidermal surface and the underlying paniculus adiposus muscle; the dermis as the layer extending from the epithelial basement membrane to the hypodermal fat layer; the epidermis as the layer from the deep border of the basal epithelial layer to the upper boundary of the cornified layer.[20] The same parameters must be used throughout a study, and any tissue section in which specific anatomic boundaries cannot be identified should be excluded.

In some organs, such as the eye, thickness measurement is complicated by normal variations in the size of the structure. As an example, the peripheral retina is less than half the thickness of the retina adjacent to the optic nerve. When measurements are made, both the location and the extent of the region measured become part of the protocol. An example of a set of guidelines for measuring retina might include: (1) a field beginning at the edge of the optic nerve and extending temporally to include all retina visible at a magnification of 100×, (2) full retinal thickness defined as extending from the internal limiting membrane to the apex of the retinal pigment epithelium, or (3) 10 thickness measurements taken in the defined field to determine mean retinal thickness. Consistent application of these limitations will assure reliable and reproducible results. Similar approaches can be applied to any organ.

## B. QUANTITATIVE MORPHOMETRICS

Phenotypic differences between mouse strains are often initially observed as variation in some morphological feature. Distinguishing putative interstrain differences from normal variation between individuals of the same strain requires some sort of quantification of the anatomical differences. Characterizing the mutant phenotype may also involve quantifying relevant morphological features. Relevant features may be at any level of organization: gross, organ, tissue, cell, or subcellular. Morphometric data can be collected in several different ways, depending on the nature of the relevant morphological features under consideration as well as the degree of variation observed.

Obvious differences in some readily measurable morphological feature may permit very straightforward analysis. Examples would be large differences in length of structures or thickness of tissue layers visualized in histological sections. Morphometric data can be collected in these instances using a measuring scale in the ocular of the microscope or with a ruler and photographic images. Standard statistical analysis can then be used to determine averages, variation, and significant differences. It may, however, be necessary to measure more complex feature parameters, such as area, perimeter, number, or lengths of curved objects. In addition, it may often be more meaningful to express these parameters in relation to the real world of three dimensions. Thus, areas become volumes, and perimeters become surface areas. This information, as well as one-dimensional data, such as number and length, are best expressed in relation to total area or volume and requires more advanced techniques. One of the most powerful methods to achieve these ends is *stereology*.

Stereology is a method for obtaining quantitative three-dimensional morphological data from two-dimensional images of sectioned material. Although a discussion of the principles and techniques of stereology is beyond the scope of this book, several reviews have been published.[22-24] The technique provides the ability to quantify tissue elements in terms of their number, length, surface area, and/or volume in relation to some reference parameter, typically the total volume of the tissue type under investigation. The basic principle of stereology is the use of standardized test grids applied to images of sectioned material. Counting the number of interactions between the test grid and the features of interest provides data that can be used in well-established equations to derive three-dimensional quantitative information in an efficient manner.

The use of complex specimen parameters such as color and density to describe and quantify morphological features requires the use of computerized image analysis. In this case, methods of specimen preparation must provide images in which the structures of interest can be readily segmented by computer-based image analysis techniques from the surrounding tissue. Such techniques may include fluorescence- or colorimetric-based immunolabeling of particular tissue elements. Standard or specialized histochemical stains may be used to selectively dye particular cell types or tissue elements. These specimens can then be analyzed using digital imaging and computer-aided segmentation. In all cases it is essential that the computer programs have facilities for operator interaction, both in terms of defining the means by which the computer selects tissue elements for analysis (such as color, intensity, or shape) and to provide for the ability to edit the data set to correct for errors both in excluding the structures of interest and including irrelevant structures.

The use of computerized image analysis to aid in obtaining morphometric data promises to be valuable in more straightforward analysis as well. Digital imaging produces images that are immediately available for analysis on the computer screen. These images can be calibrated to real dimensions, and data can be obtained quite efficiently, using the computer mouse as a "digital ruler" to define the boundaries of the tissue elements to be measured. These data can be readily tabulated for analysis by spreadsheet or statistical software. An example would be analysis of skin sections.[21] In this case, the computer mouse was used to rapidly measure the thickness of the various skin layers. More complex measurements, such as areas or proportional area, can also be generated for either user-defined or computer-segmented regions.

The kinetic and morphometric techniques described in this chapter are useful for quantitating growth and development processes as well as for documenting differences between control and experimental animals. Applications of these procedures can provide comparative numerical data useful for statistical analysis.

## ACKNOWLEDGMENTS

The authors thank Jennifer Smith and Beth Sundberg for preparation of the graphics, and Dr. Sundberg for critical review and editing of this manuscript.

This work was supported by grants from the National Institutes of Health (CA34196, AR43801, and RR8911) and from PXE International.

## REFERENCES

1. Smith, R. S., Hawes, N. L., Kuhlmann, S. D., Heckenlively, J. R., Chang, B., Roderick, T. H., and Sundberg, J. P., *Corn1*: A mouse model for corneal surface disease and neovascularization, *Invest. Ophthal. Vis. Sci.*, 37, 397, 1996.

2. Scragg, M. A. and Johnson, N. W., Epithelial cell kinetics, *J. Oral Pathol*, 11, 102, 1982.

3. Wei, Z., Cotsarelis, G., Sun, T. T., Lavker, R. M., Label-retaining cells are preferentially located in fornical epithelium: Implications on conjunctival epithelial homeostasis, *Invest. Ophthalmol. Vis. Sci.*, 36, 236, 1995.

4. Cotsarelis, G., Sun, T. T., and Lavker, R. M., Label-retaining cells reside in the bulge area of pilosebaceous unit: Implications for follicular stem cells, hair cycle, and skin carcinogenesis, *Cell*, 61, 1329, 1990.

5. Sundberg, J. P., Dunstan, R. W., Roop, D. R., and Beamer, W. G., Full-thickness skin grafts from flaky skin mice to nude mice: Maintenance of the psoriasiform phenotype, *J. Invest. Derm.*, 102, 781, 1994.

6. del Rio, J. A. and Soriano, E., Immunocytochemical detection of 5'-bromodeoxyuridine incorporation in the central nervous system of the mouse, *Exp. Brain Res.*, 49, 311, 1989.

7. Gao, C. Y. and Zelenka, P., Cyclins, cyclin-dependent kinases and differentiation. *Bioessays*, 19, 307, 1997.

8. Galand, P. and Degraef, C., Cyclin/PCNA immunostaining as an alternative to tritiated thymidine pulse labeling for marking S phase cells in paraffin sections from animal and human tissues, *Cell Tissue Kinet.*, 22, 383, 1989.

9. Hofstadter, F., Knuchek, R., and Ruschoff, J., Cell proliferation assessment in oncology, *Virchows Archiv.*, 427, 323, 1995.

10. Ghazvini, S., Kroll, S., Char, D. H., and Frigillana, H., Comparative analysis of proliferating cell nuclear antigen, bromodeoxyuridine, and mitotic index in uveal melanoma. *Invest. Ophthalmol. Vis. Sci.*, 36, 2762, 1995.

11. Greenwell, A., Foley, J. F., and Maronpot, R. R., An enhancement method for immunohistochemical staining of proliferating cell nuclear antigen in archival rodent tissues, *Cancer Letters*, 59, 251, 1991.

12. Greenwell, A., Foley, J. F., and Maronpot, R. R., Detecting proliferating cell nuclear antigen in archival rodent tissues. *Env. Health Perspect.*, 101(Suppl. 5), 207, 1991.

13. Gerdes, J., Li, L., Schlueter, C., Duchrow, M., Wohlenberg, C., Gerlach, C., Stahmer, I., Kloth, S., Brandt, E., Flad, H. D., Immunobiochemical and molecular biological characterization of the cell proliferation-associated nuclear antigen that is defined by the monoclonal antibody Ki-67, *Am. J. Pathol.*, 138, 867, 1991.

14. Abele, M. C., Valente, G., Kerim, S., Navone, R., Onesti, P., Chiusa, L., Resegotti, L., and Palestro, G., Significance of cell proliferation index in assessing histological prognostic categories in Hodgkin's disease, *Haematologica*, 82, 281, 1997.

15. Plumb, J. A. and Wright, N. A., Epidermal cell population kinetics, in *Methods in Skin Research*, Skerrow, D. and Skerrow, C. J., Eds., John Wiley & Sons, New York, 1985.

16. Sundberg, J. P., Boggess, D., Hogan, M. E., Sundberg, B. E., Rourk, M. H., Harris, B., Johnson, K., Dunstan, R. W., and Davisson, M. T., Harlequin icthyosis (*ichq*): A juvenile lethal mouse mutation with icthyosiform dermatitis, *Am. J. Pathol.*, 151, 293, 1997.

17. Basgen, J. M., Nevins, T. E., and Michael, A. F., Quantitation of antigen in tissue by immunofluorescence image analysis, *J. Immunol. Meth.*, 124, 77, 1989.

18. Green, C. R., Peters, N. S., Gourdie, R. G., Rothery, S., and Severs, N. J., Validation of immunohistochemical quantification in confocal scanning laser microscopy: A comparative assessment of gap junction size with confocal and ultrastructural techniques. *J. Histochem. Cytochem.*, 41, 1339, 1993.

19. Good, M. J., Hage, W. J., Mummery, C. L., De Laat, S. W., and Boonstra, J., Localization and quantification of epidermal growth factor receptors on single cells by confocal laser scanning microscopy, *J. Histochem. Cytochem.*, 40, 1353, 1992.

20. McAlarney, M. E., Use of the boundary element method in morphometrics, *J. Biomech.*, 28, 609, 1995.

21. Sundberg, J. P., Rourk, M. H., Boggess, D., Hogan, M. E., Sundberg, B. A., and Bertolino, A. P., Angora mouse mutation: Altered hair cycle, follicular dystrophy, phenotypic maintenance of skin grafts, and changes in keratin expression, *Vet. Pathol.*, 34, 171, 1997.

22. Gundersen, H., Bendtsen, T. F., Korbo, L., Marcussen, N., Moller, A., Nielsen, K., Nyengaard, J. R., Pakkenberg, B., Sorenson, F. B., Vesterby, A., and West, M. J., Some new, simple and efficient stereological methods and their use in pathological research and diagnosis, *APMIS.*, 96, 379, 1988.

23. Mayhew, T. M., A review of recent advances in stereology for quantifying neural structure, *J. Neurocytol.*, 21, 313, 1992.

24. Bolender, R. P., Hyde, D. M., and Dehoff, R. T., Lung morphometry: A new generation of tools and experiments for organ, tissue, call, and molecular biology, *Am. J. Physiol.*, 265, L521, 1993.

# 9 Ultrastructural Evaluation of Mouse Mutations

*Lesley S. Bechtold*

## CONTENTS

## I. INTRODUCTION

Scanning (SEM) and transmission (TEM) electron microscopy provide valuable tools for assessing morphological changes found in various mouse mutations. The theory behind the preparation of samples for both types of EM is the same: samples are collected and fixed (if necessary) in such a way that the morphology of the finished product is as close to lifelike as possible. Consequently, the speed with which the samples are collected and fixed in a carefully selected fixative and buffering system all determine the quality of the final result. Care and attention to detail provide high-quality results that support other studies.

Electron microscopic evaluation of tissues provides the investigator with an added dimension to the interpretation of normal and pathological conditions in the mice being investigated. Scanning electron microscopy provides magnification comparable and slightly higher than that obtained with a light microscope. Although images obtained are not in color, they provide a three-dimensional view of the tissue. This helps correlate two-dimensional observations of routine hematoxylin and eosin stained histologic slides with what is actually going on in real life. For example, hair fibers surrounded by cornified debris within the follicular infundibulum in histologic sections were observed to form compact sheaths around emerging hairs when viewed by scanning electron microscopy for the mouse mutation called harlequin ichthyosis.[1] Modifications of this technique, such as freeze fracture, can provide three-dimensional views of "cut" edges of tissues and surface texture.

Transmission electron microscopy provides fine detail at extremely high resolution of two-dimensional images of tissues. Mouse kidneys can be examined to determine if the amorphous, eosinophilic material within glomeruli, as seen by light microscopy, is composed of amyloid proteins, based on ultrastructural features.[2]

The various electron microscopic methods are expensive and labor intensive. Therefore, this approach usually serves as an adjunct to light microscopy and histopathology, rather than a primary method. Histopathology is useful to determine which structures should be examined and which specific anatomic region of the organ should be the focus of attention. As with all experiments, controls (age- and sex-matched wild type, +/+, mice) should be examined concurrently. Images with matching magnification of mutant and control anatomical structures can provide dramatic and convincing evidence of altered phenotype.

This chapter focuses on specialized methods for tissue collection and processing to provide specimens suitable for examination in either a transmission or scanning electron microscope.

## II. SPECIMEN COLLECTION AND FIXATION

There are two basic methods of fixation: immersion fixation and perfusion. During immersion fixation, the specimen is immersed in the fixative and the fixative penetrates the cells of the specimen. Because the rate of penetration of fixatives tends to be slow, the pieces should be as small as possible, 1 mm$^3$ or less, and the fixation time can be fairly long. Alternatively, for tissues such as skin, very thin slices or strips may also be fixed in this manner as long as they are 1 mm thick or less. This method of fixation also works very well for small, single cells such as red blood cells and sperm. Tissue that is collected for EM during necropsy must be minced or sliced in a large drop of fixative with an ethanol-cleaned razor and placed in a vial of cold fixative as quickly as possible to minimize shrinkage artifact during fixation. The razor blades used to mince tissue are wiped clean with a tissue wetted with ethanol to prevent the small metal filings on the edge of the razor blade from becoming embedded in the tissue being prepared for EM. If these particles do become embedded in the tissue, they can damage the edge of the knife during sectioning.

Perfusion is used when larger specimens, whole organs, such as the brain, or even whole animals need to be collected. Transcardial perfusion allows the fixative to be circulated throughout the entire body of an animal and fully penetrate large organs which would remain largely unfixed if removed whole and immersion-fixed. The circulatory system is flushed first with 0.01 $M$ phosphate-buffered saline (PBS) to remove as much of the blood as possible. Then fixative is perfused throughout the animal, slowly or quickly depending on the fixative used. Once all the fixative has been used, specimens are removed and placed in additional fixative for three hours to overnight at 4°C to allow additional fixation to occur. More details on this method are provided in Chapter 5, Necropsy Methods for Laboratory Mice.

Some specimens, such as hair and nails, do not require any fixation because they do not contain water and are not damaged by changes in pH and osmolarity. They can also be collected at the time of necropsy if the animal is to be sacrificed, or they can be collected while the animal is alive, since clipping hair and nails does no harm to the animal. Fixation and the subsequent dehydration are bypassed during preparation for EM.

## III. FIXATIVES

Fixatives are used to preserve the natural morphology and ultrastructure of a specimen. The most widely used fixatives for EM are the noncoagulant fixatives, such as glutaraldehyde, paraformaldehyde, acrolein, and osmium tetroxide. They are known as noncoagulant fixatives because they crosslink proteins and make them less likely to become extracted by solvents during processing, but they do not coagulate or denature the proteins. They are also called additive fixatives because they work by adding themselves onto the cellular proteins and stabilizing the ultrastructure of the specimen.

Glutaraldehyde is the most commonly used fixative for EM and is generally used in conjunction with one or more of the other fixatives during processing. Because glutaraldehyde tends to penetrate very slowly during perfusion, it is frequently used in a glutaraldehyde–paraformaldehyde mix known as Karnovsky's Fix.[3] Karnovsky's Fixative is usually made up in phosphate buffer and may contain up to 4% glutaraldehyde and 5% paraformaldehyde. The concentration of each varies according to the preference of the person preparing the specimen. Paraformaldehyde by itself is not suitable as a primary fixative for EM. Paraformaldehyde penetrates much more rapidly than glutaraldehyde, but its effects are reversible and it can be washed out. Consequently a mixture of the two during perfusion followed by immersion fixation is recommended. The effects of glutaraldehyde are not reversible.

Acrolein, a precursor of glutaraldehyde, can also be used as a primary fixative in place of glutaraldehyde. It tends to penetrate very rapidly with nonreversible effects. Consequently, it can be used when fixation must be very rapid. It can also be used in a mixture with paraformaldehyde with excellent results.

Heavy metal fixatives, such as osmium tetroxide, penetrate very slowly but because of their "additive" effect to cell membranes, they impart superb contrast to TEM specimens. These fixatives tend to be used as secondary fixatives and are used after the tissue has been fixed with an aldehyde fixative. Osmium and glutaraldehyde will react with each other, so they are used separately, and the specimen is well washed with buffer in between fixatives. Ruthenium can be used in place of osmium with similar results.[4]

In general, fixatives are used at 4°C. This will slow down the rate of penetration, but it will also slow down extraction and shrinkage. It is preferable to make fresh fixative solutions because the aldehyde fixatives tend to acidify after several days. Fixatives are usually made by purchasing small volumes of distilled, concentrated fixative which can be diluted to the desired concentration in a buffer. Osmium tetroxide can be purchased as an aqueous solution or as a pure crystal. If purchased as a crystal, it must be prepared by dissolving it in distilled water at least a day in advance because it takes several hours to dissolve. The aqueous solution, which can be kept at 4°C indefinitely, is then diluted in buffer as needed. It is also important to remember that all fixatives are toxic and should be used in a fumehood while wearing gloves for protection.

## IV. BUFFERING SYSTEMS

There are several buffers available that can be used for specimen preparation. Sorenson's phosphate buffer[5] is the most widely used buffering system in EM for several reasons. It is considered a physiological buffer because it is very similar to the buffering system found in animals and produces the most natural conditions during fixation. It also tends to produce fewer artifacts and fixative precipitate overall. It maintains its pH during fixation and works well in the physiological range of 6.8 to 7.4. Unlike other buffers, such as collidine and cacodylate, phosphate buffer is nontoxic.

Like fixatives, phosphate buffer is also better when fresh. After several weeks, it can become contaminated and provide an environment for microorganisms. Any phosphate buffer that is older than one month should be discarded and fresh buffer made. It is generally made by combining 0.2 $M$ monobasic and dibasic phosphate solutions and diluting to produce the desired final concentration, 0.1 $M$ being the most commonly used molarity. Apart from providing a vehicle for the fixative, buffer is also used to wash the specimen after each fixation step to remove residual fixative from the specimen.

Cacodylate buffer is another buffer of choice for EM. Cacodylate buffer contains arsenic, but unlike phosphate buffer, it can be stored for several months. Cacodylate can be used interchangeably with phosphate. Both have excellent buffering capacity and provide comparable results. Cacodylate buffer is made by combining 0.2 $M$ cacodylate solution and 0.2 $M$ HCl and diluting to produce a final concentration of 0.1 $M$.

## V. DEHYDRATION

The purpose of dehydrating specimens is to gradually but completely remove any water present in the sample and replace it with a liquid that is either miscible with the resin that the specimen will ultimately be embedded in or will evaporate and leave the specimen completely desiccated.

For TEM, any solvent such as ethanol, methanol, or acetone may be used. The most commonly used solvent is ethanol, but others are used based on the user's preference. There are benefits and disadvantages to each. Acetone tends to take up water more easily and is more toxic than ethanol, but it does not require a transition fluid to get the specimens into resin. Acetonitrile may also be used during dehydration when lipid extraction is a concern.[6] Specimens are transferred through a graded series of the solvent, starting at 50 to 60% solvent in distilled water. Several changes of 100% solvent will ensure that all the water has been removed from the specimen and replaced with solvent. If ethanol is used, it is advisable to use a couple of quick changes in a transfer fluid such as propylene oxide. Propylene oxide tends to be more miscible than most solvents in resin, but as it also extracts cellular components rapidly, it is not used for the entire dehydration process.

For SEM, the same dehydration schedule is used to replace the water in the specimen with ethanol. It is recommended that a graded series beginning at 25 to 40% ethanol be used to remove all traces of buffer and avoid any problems with precipitation. The specimen is gradually transferred into liquid $CO_2$, rather than resin, which becomes gaseous $CO_2$ during the process of critical point drying. Critical point drying replaces all liquid in the specimen with $CO_2$, leaving the specimen completely desiccated. The critical point drying should be done as slowly and gently as possible so the specimen is not damaged in the process. For very delicate specimens, it is possible to bypass critical point drying and instead use a transition fluid, such as hexamethyldisilazane[7] or dimethoxypropane,[8] which can be evaporated in a fume hood, leaving the specimen completely dry. Once the specimens for SEM are dry, they are ready for coating.

As with fixatives and buffers, the volume of solvent used during each step of the dehydration process should be much greater than that of the specimen. It is also advisable to use freshly opened bottles of 100% solvent during the final dehydration steps. Dehydration is generally done at room temperature and changes of 10 to 15 minutes are usually adequate for most specimens.

## VI. EMBEDDING

Epoxy resins are the most widely used resins for several reasons: they are readily available, the components have a long shelf life, their penetration into tissue is very good, the sections are stable under the electron beam, and they provide consistent results. Their only disadvantage is that they are not water miscible. Consequently, specimens need to be dehydrated before they can become permeated with resin. There are several kinds of epoxy resins that are used: Embed (formerly known as Epon), Araldite, Spurr's, and combinations such as Embed–Araldite.[9] All are used according to the user's preference. Once the specimen has been completely infiltrated with resin over several hours or days, the resin is heat polymerized, and the resulting blocks of hardened resin containing the specimen are ready for sectioning and staining.

## VII. HEAVY METALS AND ELECTRON MICROSCOPY

As a science, electron microscopy relies primarily on heavy metals to make specimens stable enough to withstand the electron beam. Without the heavy metals used during preparation, delicate biological specimens would be burned up under the intensity of the electrons. Heavy metal fixatives insert themselves into the membranes, providing stability to the entire structure. Heavy metal stains, such as uranyl acetate and lead citrate,[10] permeate ultrathin sections and are absorbed by the tissue, thus stabilizing the sections. In SEM, the thin layer of gold, gold–palladium, platinum, or carbon protects the specimen from the electron beam. Without these, the resolution achieved with electron microscopy would not be possible.

## VIII. ROUTINE PROTOCOLS

Routine protocols for preparing specimens are listed below. Those used for TEM will then require sectioning and staining before they can be evaluated. Upon completion of the SEM protocols, specimens can be examined with an appropriate microscope. All materials listed below can be obtained from Electron Microscopy Sciences, Ft. Washington, PA.

### A. STANDARD PROTOCOL FOR TEM SPECIMEN PREPARATION

1. Fixation—2% glutaraldehyde and 2% paraformaldehyde in 0.1 $M$ phosphate buffer, pH 7.2. Fix overnight at 4°C.
2. Buffer Washes—Wash specimen twice for 15 minutes with phosphate buffer at 4°C.
3. Postfixation—1% osmium tetroxide in phosphate buffer. Fix overnight at 4°C.
4. Buffer Washes—Wash specimen twice for 15 minutes with phosphate buffer at 4°C.
5. Dehydration—60% ethanol - 15 minutes at room temperature.
   80% ethanol—15 minutes at room temperature.
   95% ethanol—15 minutes at room temperature.
   100% ethanol—twice for 15 minutes at room temperature.
6. Transition Fluid—Propylene oxide—twice for 10 minutes at room temperature.
7. Infiltration—Propylene oxide: resin (1:1)—overnight on rotator at room temperature.
   Pure Resin—24 hours on rotator at room temperature.

Prepare fresh resin and embed specimens in molds. Cure at 65°C for 72 hours.

### B. STANDARD PROTOCOL FOR SEM SPECIMEN PREPARATION

1. Fixation—2% glutaraldehyde and 2% paraformaldehyde in 0.1 $M$ phosphate buffer, pH 7.2. Fix for 4 hours at 4°C.
2. Buffer Washes—Wash three times for 15 minutes in phosphate buffer at 4°C.
3. Postfixation—Postfix for 2 to 3 hours in 1% osmium tetroxide in phosphate buffer at 4°C.
4. Buffer Washes—Wash three times for 15 minutes with phosphate buffer at 4°C.

5. Critical Point Dry—Flush specimen gently for 5 minutes four times with $CO_2$. Gradually increase temperature in the critical point drier to 41°C. Over a period of 30 to 40 minutes, release pressure slowly to allow $CO_2$ to evaporate, while allowing temperature to return to room temperature.
6. Coating—Use double-sided tape to glue the specimen onto the surface of a clean aluminum SEM stub. Coat specimen in sputter coater with a 15 nm coating of gold.

*Note*—Steps 3 and 4 are optional for SEM specimen preparation, and steps 1 through 4 are unnecessary for hair and nail specimens. Fixed, single-cell suspensions, such as red blood cells and sperm, will adhere to 13 mm circular coverslips which have been treated with 1% poly-L-lysine for 5 minutes. The coverslips are then carried through all the remaining steps of preparation.

## C. Phosphate Buffer, pH 7.2

Solution A— 0.2 $M$ sodium phosphate, monobasic: dissolve 27.6 g of $NaH_2PO_4 \cdot H_2O$ in 1000 ml of distilled water.

Solution B—0.2 $M$ sodium phosphate, dibasic: dissolve 35.61 g of $Na_2HPO_4 \cdot 2H_2O$ in 1000 ml of distilled water.

Prepare 2000 ml of 0.1 $M$ buffer by mixing 280 ml of solution A with 720 ml of solution B and 1000 ml of distilled water. Check pH of buffer and adjust if necessary with 1 N NaOH or 1 $M$ HCl.

## D. Cacodylate Buffer, pH 7.2

Solution A—0.2 $M$ sodium cacodylate: dissolve 42.8 g of cacodylic acid in 1000 ml of distilled water.

Solution B—0.2 $M$ HCl – dilute 10 ml of concentrated (36 to 38%) HCl in 603 ml of distilled water.

Prepare 200 ml of 0.1 $M$ buffer by adding 4.2 ml of solution B to 50 ml of solution A and diluting to 200 ml with distilled water.

*Note*—Use undiluted, 0.2 $M$ buffer to prepare fixatives and replace part of the distilled water with concentrated fixative to achieve the desired final concentration of the fixative.

## E. Epon–Araldite Resin

Using a disposable 100 ml plastic tripour beaker, combine the following ingredients on a stir-plate at room temperature for at least 30 minutes:

14 ml DDSA (dodecenyl succinic anhydride)
6 ml Embed 812
4 ml Araldite 506
1 ml DBP (dibutyl phthalate)
0.7 ml DMP-30 (accelerator)

This resin can be mixed with propylene oxide during infiltration and used full strength to embed specimens. It will polymerize at 65°C.

## F. URANYL ACETATE STAIN

Dissolve 1 g of uranyl acetate in 50 ml of distilled water. Stain grids for 1 to 2 hours using this stain.

## G. LEAD CITRATE STAIN

Dissolve 1.33 g of lead nitrate and 1.76 g of sodium citrate in 30 ml of distilled water and mix well on a stir plate for 30 minutes. Add 10 ml of 1 N NaOH. Mix for an additional 5 minutes. Add 10 ml of distilled water, mix and store at 4°C until used. Stain grids for 1 hour after staining with uranyl acetate and washing between stains with distilled water.

## ACKNOWLEDGMENTS

This work was supported by a grant from the National Cancer Institute (CA34196).

## REFERENCES

1. Sundberg, J. P., Boggess, D., Hogan, M. E., Sundberg, B. A., Rourk, M. H., Harris, B., Johnson, K., Dunstan, R. W., and Davisson, M. T., Harlequin ichthyosis. A juvenile lethal mouse mutation with ichthyosiform dermatitis, *Am. J. Pathol.*, 151, 293, 1997.
2. Sundberg, J. P., France, M., Boggess, D., Sundberg, B. A., Beamer, W. G., and Shultz, L. D., Development and progression of psoriasiform dermatitis and systemic lesions in the flaky skin (*fsn*) mouse mutation, *Pathobiology*, 65, 261, 1997.
3. Karnovsky, M. J., A formaldehyde-glutaraldehyde fixative of high osmolality for use in electron microscopy, *J. Cell. Biol.* 27, 137A, 1965.
4. Swartzendruber, D., Burnett, I., Wertz, P., Madison, K., and Squirer, C., Osmium tetroxide and ruthenium tetroxide are complementary reagents for the preparation of epidermal samples for transmission electron microscopy, *J. Invest. Dermatol.*, 104, 417, 1995.
5. Dawson, R. M. C., Elliott, D. C., Elliott, W. H., and Jones, K. M., *Data for Biochemical Research*, 2nd ed., Clarendon Press, Oxford, 1969.
6. Edwards, H. H., Yeh, Y.-Y., Tarnowski, B. I., and Schonbaum, G. R., Acetonitrile as a substitute for ethanol/propylene oxide in tissue processing for transmission electron microscopy: comparison of fine structure and lipid solubility in mouse liver, kidney and intestine, *Microscopy Res. Tech.*, 21, 39, 1992.

7. Bray, D. F., Bagu, J., and Koegler, P., Comparison of hexamethyldisilazane (HMDS), Peldri II, and critical-point drying methods for scanning electron microscopy of biological specimens, *Microscopy Res. Tech.*, 26, 489, 1993.
8. Weyda, F., Simple desiccation method for scanning electron microscopy using dimethoxypropane, *Proc. 50th An. Meeting Electron Microscopy Soc. Am.*, San Francisco Press, California, 1992.
9. Mollenhauer, H. H., Plastic embedding mixtures for use in electron microscopy, *Stain Tech.*, 39, 111, 1964.
10. Reynolds, E. S., The use of lead citrate at high pH as an electron opaque stain in electron microscopy, *J. Cell Biol.*, 17, 208, 1963.

## SUGGESTED READING

1. Glauert, A. M., Fixation, *Dehydration and Embedding of Biological Specimens*, Elsevier North-Holland Biomedical Press, Amsterdam, The Netherlands, 1975.
2. Hayat, M. A., *Fixation for Electron Microscopy*, Academic Press, New York, 1981.
3. Hayat, M. A., *Principles and Techniques of Electron Microscopy, Biological Applications*, 3rd edition, CRC Press, Boca Raton, 1989.

# 10 Immunohistochemical and Immunofluorescence Methods

*Melissa J. Relyea, John P. Sundberg, and Jerrold M. Ward*

## CONTENTS

## I. INTRODUCTION

Immunofluorescence and immunohistochemistry are two methods with similar procedures that are commonly used to evaluate expression of proteins in tissues. Their value is that expression can be identified in specific cells, organs, and structures, as opposed to immunoblots, which merely indicate whether expression is present or not in an organ. Immunofluorescence and immunohistochemistry are qualitative, while immunoblots can be quantitative or semiquantitative, so each does provide advantages.

Immunofluorescence is a rapid and accurate method for demonstrating protein expression in tissues. It is a very sensitive method because frozen specimens are used, avoiding alteration of the tertiary structures of the proteins by fixation. The structural problems evident when using frozen sections with enzyme systems are minimal with immunofluorescence because the unstained background is nearly invisible. However, evaluation of fluorescently labeled tissues requires a fluorescence microscope, and photography of images dictates the use of very fast films (grainy images can be a problem) in a vibration-free environment. Fluorescently labeled preparations are not permanent. The labels may fade in the light and are not compatible with permanent organic-solvent-based mounting media. These latter features limit the value of this technique for general purposes.

Immunohistochemistry has evolved to circumvent many of the problems found with immunofluorescence. Paraffin-embedded sections can be used in many instances to provide high-quality images that can be viewed by light microscopy. Photomicrographs can be taken with slow, low-grain films. With adequate controls, correct expression data can be obtained. Unfortunately, some antibodies work well only with frozen specimens, such as many of the mouse lymphocyte markers. Frozen sections can be processed immunohistochemically to yield slides that can be viewed with a light microscope. Unfortunately, architecture is poorly preserved in frozen sections, so obtaining reproducibly high-quality specimens of publication caliber can be very difficult.

Collectively, immunofluorescence and immunohistochemistry have become standard tools for pathologists in evaluating diagnostic specimens. The approaches are applicable to experimental studies as well.

## II. HISTORY

In the early 1940s, Albert H. Coons and his colleagues were the first to identify a specific tissue antigen using a fluorescence microscope and an antibody labeled with a fluorescent dye.[1-3] As a result of their work, histopathology became a much more accurate science, whereas the diagnostic staining techniques of the past were often open to somewhat subjective interpretations.

Although fluorescein isocyanate was the first fluorescent dye to be used in an immunohistochemical reaction, fluorescein isothiocyanate (FITC) soon became the most commonly utilized label because of its stability and ease of conjugation. As time passed, different fluorophores of various colors were developed, and experiments began with enzyme-labeling systems that allowed one to view the reaction under a conventional light microscope, circumventing many of the problems inherent to immunofluorescence.

Horseradish peroxidase was the first enzyme to be used in an immunohistochemical procedure.[4,5] Development of other enzyme systems followed, including alkaline phosphatase,[6] glucose oxidase,[7] and $\beta$-D-galactosidase.[8] All of these are discussed in the section on enzyme systems.

Several immunohistochemical methods evolved after the direct labeled antibody technique. These included the labeled secondary antibody procedure, which provided a small degree of amplified signal, the peroxidase–anti-peroxidase (PAP) technique,[9] which further increased signal amplification, and the avidin–biotin system,[10,11] which provides the greatest degree of amplification and is the most common immunohistochemical method used today.

## III. ANTIBODY SOURCES

The success of any immunohistochemical procedure depends upon high-quality antibodies and reagents. Obtaining antibodies that work well in mouse tissues can be difficult. For example, hormones produced in the pars distalis of the pituitary are functionally similar in all mammals, but their epitopes are, unfortunately, species specific. Some companies provide data on cross reactivity between species or will provide antibodies with a return policy if they do not work. A list of some of the vendors is provided in Table 10.1. The American Type Culture Collection (ATCC) provides large numbers of antibodies and hybridoma lines to produce one's own monoclonal antibodies, as well. Individual investigators are frequently a very good resource for high-quality, concentrated, and well-defined antibodies. They will often provide small aliquots of their own reagents when asked, until demand gets too high and then they choose to provide them through commercial vendors.

## IV. IMMUNOHISTOCHEMISTRY

### A. BASIC CONCEPTS

Before beginning any immunohistochemical procedure, the tissues to be used must be prepared properly. Paraffin-embedded tissue must be deparaffinized in xylene and rehydrated through a series of graded alcohols to water.    Slides cut from frozen tissue should be air dried at room temperature for at least one hour, or up to overnight, and should be fixed briefly before processing (see Detailed Protocols.) Slides with frozen sections may be stored at -20°C or below (preferably -80°C) after they are dried. They should be wrapped individually in aluminum foil or plastic wrap and sealed in a moisture-free, air-tight container with a desiccant.    Bring them to room temperature within the container before use, to prevent condensation.

   The first step of the immunohistochemical reaction is the incubation of the hydrated tissue in a primary antibody raised against a specific antigen, such as a rabbit-raised anti-papillomavirus. If the primary antibody is not biotinylated, a secondary biotinylated antibody is applied, which in this case would be biotinylated anti-rabbit IgG, which binds to the primary antibody and introduces biotin molecules to the area. The tissue is then incubated in an avidin–biotinylated enzyme complex, which binds to the biotin molecules on the secondary antibody. Next, a chromogenic substrate for the enzyme is applied, producing color deposition in the positive areas. Finally, a counterstain is chosen that is compatible with the selected substrate, and the tissue is counterstained and mounted in the proper mounting medium.

   Endogenous peroxidase activity can be minimized by incubation of slides in 3% $H_2O_2$ in methanol for 15 to 30 minutes before antibodies are added. To block nonspecific staining, initial incubation of the tissue in normal serum from the host of the secondary antibody may be necessary. Ova-albumin or bovine serum albumin may also be used to block nonspecific staining, but if you are using a commercially available kit, the normal serum provided often yields optimal results.

## TABLE 10.1 VENDERS THAT PROVIDE ANTIBODIES AND KITS

**American Type Culture Collection (ATCC)**, 12301 Parklawn Drive, Rockville, MD 20852,
   Tel: 800/638-6597, Fax: 301/816-4379, E-mail: tech@atcc.org, www.atcc.org
**Antibody Resource Page Online**, http://antibodyresource.com/onlinecomp.html
**Berkeley Antibody Company (BAbCO)**, 1223 South 47th Street, Richmond, CA 94804,
   Tel: 800/92-BABCO, Fax: 510/412-8940, E-mail: product@babco.com, www.babco.com
**The Binding Site Inc.**, 5889 Oberlin Drive, Suite 101, San Diego, CA 92121,
   Tel: 800/633-4484, Fax: 619/453-9189
**Biodesign International**, 105 York Street, Kennebunk, ME 04043,
   Tel: 207/985-1944, Fax: 207/985-6322, E-Mail: info@biodesign.com, www.biodesign.com
**Biomedia**, P.O. Box 8045, Foster City, CA 94404,
   Tel: 800/341-8787, Fax: 415/341-2299, www.biomeda.com
**Caltag Laboratories**,1849 Old Bayshore Highway, Suite 200, Burlingame, CA 94010,
   Tel: 800/874-4007, Fax: 650/652-9030, www.caltag.com
**Chemicon International, Inc.**, 28835 Single Oak Drive, Temecula, CA 92590 ,
   Tel: 800/437-7502, Fax: 909/676-9209, www.chemicon.com
**DAKO Corporation**, 6392 Via Real, Carpenteria, CA 93013,
   Tel: 800/235-5763, Fax: 805/566-6688, www.dakousa.com/dakousa
**Immunotech, Inc.**, 160B Larrabee Rd., Westbrook, ME 04092,
Tel: 800/458-5060, Fax: 207/854-0116, www.immunotech.com
**Jackson ImmunoResearch Laboratories Inc.**,
   872 West Baltimore Pike, P.O. Box 9, West Grove, PA 19390,
   Tel: 800/367-5296, Fax: 610/869-0171, www.jacksonimmuno.com
**Kirkegaard & Perry Laboratories, Inc.**, 2 Cessna Court, Gaithersburg, MD 20879,
   Tel: 800/638-3167, Fax: 301/948-0169, www.kpl.com
**Linscotts Directory**, Linscotts, USA, 4877 Grange Road, Santa Rosa, CA 95404,
   Tel: 707/544-9555, Fax: 415/389-6025,
   http://ourworld.compuserve.com/homepages/LINSCOTTSDIRECTORY/
**NEN Life Science Products**, 549 Albany Street, Boston, MA 02118,
   Tel: 800/551-2121, www.nenlifesci.com
**Peninsula Laboratories, Inc.**, 611 Taylor Way, Belmont, CA 94002,
   Tel: 800/922-1516, Fax: 415/595-4071, E-mail: info@penlabs.com, www.penlabs.com
**PharMingen**, 10975 Torreyana Road, San Diego, CA 92121,
   Tel: 800/848-6227, Fax: 619/812-8888, www.pharmingen.com
**QED Advanced Research Technologies**, 11011 Via Frontera, San Diego, CA 92127,
   Tel: 800/929-2114, Fax: 619/592-1509
**Santa Cruz Biotechnology, Inc.**, 2161 Delaware Avenue, Santa Cruz, CA 95060,
   Tel: 800/457-3801, Fax: 831/457-3801, www.scbt.com
**Serotech Inc.**, Partners 1, 1017 Main Campus Drive, Suite 2450,
   North Carolina State University, Raleigh, NC 27606,
   Tel: 800/265-7376, Fax: 919/515-9980, www.serotec.co.uk
**Sigma Chemical Company**, P.O. Box 14508, St. Louis, MO 63178,
   Tel: 800/325-3010, Fax: 800/325-5052, E-mail: sigma.biotech@sial.com
**Signet Laboratories, Inc.**, 180 Rustcraft Rd., Dedham, MA 02026,
   Tel: 800/223-0796, Fax: 617/461-2456
**Vector Laboratories, Inc.**, 30 Ingold Road, Burlingame, CA 94010,
   Tel: 800/227-6666, Fax: 415/697-0339, www.vectorlabs.com
**Zymed Laboratories Inc.**, 458 Carlton Court, South San Francisco, CA 94080,
   Tel: 800/874-4494, Fax: 415/871-4499, E-mail: tech@zymed.com, www.zymed.com

## B. Control Tissues

With each immunohistochemical run, a control tissue should be included to help determine the success of the reaction (see Table 10.2) A known positive control is used to prove that the reaction worked. It may be a normal tissue known to express a certain antigen or an abnormal tissue already proven to react positively with the applied antibody. A normal control of the same tissue type as that being tested may be useful as a basis for comparison to highlight abnormal expression patterns in the test tissue. A bank of paraffin blocks and slides for use as normal control tissues may be created by gathering a group of age- and sex-matched inbred mice of the same strain used in your study. Collect all tissues in your usual fixative or perfuse each mouse with a different fixative for comparison studies. Fresh tissue from age-, sex-, and strain-matched mice may be frozen in O.T.C. Compound (Tissue-Tek, Sakura Finetek U.S.A. Inc., Torrance, CA) for use as controls as well. Fixatives will change the tertiary structure of proteins, thereby altering expression. When first using antibodies with which you are unfamiliar, it is useful to compare expression by immunofluorescence on frozen sections to immunohistochemistry done on the same tissue preserved in various fixatives (see Chapter 5) to determine the optimal approach.

---

**TABLE 10.2**
**Controls for Immunohistochemistry**

1. Positive Controls
   A. Tissues/cells with known strong positive immunoreactivity
      1. On same slide as test tissue/cell
      2. On different slide than test tissue/cell
2. Negative Controls
   A. Tissues with known negative immunoreactivity
      1. On same slide as test tissue/cell
      2. On different slide than test tissue/cell
3. Adsorption Controls
   A. Adsorb antibody with purified antigen/peptide
4. Background Controls
   A. Increase concentration of a control antibody or isotypic immunoglobulin or another specific antibody that is not known to immunoreact with any tissue/cell on the slide of interest to produce background/nonspecific staining
   B. Compare pattern of nonspecific staining with pattern of "specific" staining
   C. Use antibody that reacts with other tissue/cell on same slide and increase concentration to produce background on other tissues/cells

---

## C. Enzyme Systems

Three different enzyme detection systems are commonly offered: horseradish peroxidase, alkaline phosphatase, and glucose oxidase. Peroxidase-based systems are sensitive, provide dense staining, and are by far the most widely used. Alkaline phosphatase systems are used when endogenous peroxidase activity is a problem, or when high sensitivity is a must. Although highly sensitive, alkaline phosphatase staining is usually less intense. Glucose oxidase is a plant enzyme, not present in animals. It is useful on the rare occasion that endogenous peroxidase or alkaline phosphatase activity in animal tissues causes severe interference with specific immunohistochemical staining, or when multiple staining is desired. The glucose oxidase system is the least sensitive of the three. Yet another enzyme, $\beta$-D-galactosidase, is used infrequently in multiple staining reactions as well. It is derived from bacteria and thus cannot be used in preparations containing bacteria.

## D. Chromagens

The chromagen is the substance that allows visualization of the site of a positive reaction. It binds to the enzyme in the avidin–biotinylated enzyme complex. Each enzyme system uses different chromagens, and many companies have developed their own versions of these to match the enzyme systems used. The standards are as follows.

Two peroxidase chromagens are diaminobenzadine (DAB) and 3-amino-9-ethylcarbazole (AEC). DAB yields a brown precipitate and is insoluble in organic solvents. It provides good contrast with most counterstains and is compatible with permanent organic-solvent-based mounting media, such as Permount (Fischer Scientific). The color of the DAB can be confused by the neophyte with melanin pigment. AEC stains red, making beautiful color slides when used with hematoxylin. However, it is alcohol soluble and thus must be mounted using aqueous-based systems. Biomedia (Foster City, CA) sells an aqueous media called Crystal/Mount, which may be postmounted with organic-solvent-based media for improved permanence. 4-Chloro-1-naphthol may also be used to develop a peroxidase reaction. It gives a blue/grey product and is alcohol soluble.

Alkaline phosphatase chromagens include 5-bromo-4-chloro-3-indolyl phosphate/nitroblue tetrazolium (BCIP/NBT), 5-bromo-4-chloro-3-indolyl phosphate/iodoblue tetrazolium (BCIP/INT), and Fast Red/Naphthol AS-Mx. Both BCIP/NBT and BCIP/INT are organic-solvent compatible. They stain navy/black and yellow/brown, respectively. Fast Red stains a nice bright pink color, but it is alcohol soluble. Many companies have developed their own organic-solvent compatible red chromagens for alkaline phosphatase, but there are no standards for these. Two examples are Vega Red (Biomedia) and Vector Red Substrate (Vector, Burlingame, CA).

Glucose oxidase substrates are all tetrazolium salts. They include nitroblue tetrazolium (NBT) (purple/blue), tetranitroblue tetrazolium (TNBT) (black), and iodoblue tetrazolium (INT) (red). Only INT is incompatible with organic solvents.

The developing substrate for β-D-galactosidase uses 5-bromo-4-chloro-indolyl-β-D-galactopyranoside (BCIG) and ferrocyanide with ferricyanide to produce a turquoise-blue reaction product.

## E. COUNTERSTAINS

A variety of counterstains can be used effectively to identify cells and structures in the tissue and provide sharp contrast to those that stain positively in the reaction.

Counterstains for immunohistochemistry include hematoxylin (blue), toluidine blue (blue), light green (green), nuclear fast red (red), and methyl blue (blue). Hematoxylin was long ago developed empirically to optimize contrast with eosin for routine histologic sections and is the most commonly used counterstain. Light green provides excellent contrast for diaminobenzidine-based immunoperoxidase reactions and yields wonderful color transparencies for lectures. Unfortunately, a green filter is used for black-and-white photography which removes the counterstain and, therefore, the background tissue contrast, making images unusable for publication.

Counterstain choice is often subjective. However, it is wise to consider the compatibility of the counterstain with both the color of the substrate and the chemical makeup of the intended mounting medium to obtain quality, archival, highly photographable slides from your reactions.

## F. ENHANCEMENT OF IMMUNOREACTIVITY

Immunoreactivity in tissues fixed by freezing is most often sufficient without further augmentation. However, it is occasionally necessary to enhance immunoreactivity in paraffin-embedded tissues. A variety of techniques have been used to achieve this. Most of these enhancement techniques increase sensitivity of the detection system so that the antibody can be diluted to a greater degree. These include specialized fixatives (e.g., methacarn, Carnoy's, Bouin's, paraformaldehyde) and reagent kits, enhancing reagent components, heat, microwave exposure, and enzyme digestion.[12] Some of these methods have been reported as antigen retrieval or unmasking techniques.

If immunostaining is not successful with an antibody, enhancement techniques may be warranted. For example, cell surface antigens in animal tissues often are better preserved with Bouin's fixative or paraformaldehyde. Keratins are more easily demonstrated in ethanol-fixed tissues and in formalin-fixed tissues after enzyme digestion. Examples of optimal fixatives and methods for commonly used antibodies are given in Tables 10.3 and 10.4.

## V. IMMUNOFLUORESCENCE

Instead of utilizing a chromagen/enzyme reaction to deposit color in the area of a positive antigen presence, immunofluorescence uses a fluorochrome-labeled antibody, which is then viewed under a special fluorescence microscope.  Today, with the prevalence of immunoenzymatic techniques, immunofluorescence is most often

**TABLE 10.3**

**Examples of Optimal Fixatives and Enhancement Procedures for Immunohistochemistry of Tissues in Paraffin-Embedded Sections**

| Antigen | Fixative for Optimal Immunoreactivity | Enhancement Method |
|---|---|---|
| Cell surface lymphocytes and other antigens | Bouin's fixative, paraformaldehyde, methacarn | Microwave pretreatment |
| Immunoglobulins | Bouin's fixative | |
| Keratins | 70% ethanol | Enzyme digestion |
| Cell cycle proteins | Various fixatives | Microwave pretreatment |
| Viral antigens | Formalin | |
| Hormones | Formalin | |

**TABLE 10.4**

**Hematopoietic Antigens Demonstrated in Paraffin Sections of Mouse Tissues**

| Antigen | Normal Cells Expressing Antigen | Reactive in Formalin-fixed Tissues | Preferred Fixative for Optimal Immunoreactivity | Mouse Tumors Immunoreactivity |
|---|---|---|---|---|
| CD3 | Various T Lymphocytes | Yes* | Bouin's, PF, B5*** | T Cell Lymphoma |
| CD45R (B220, Ly-5) | Various B Lymphocytes | Yes* | Bouin's, PF, B5*** | B Cell Lymphoma |
| Immunoglobulins (IgG, IgA, etc.) | Immunoglobulin Producing B Cells | Yes* | Bouin's, PF, B5*** | B Cell Lymphoma |
| Immunoglobulin kappa light chains | Immunoglobulin Producing B Cells | Yes* | Bouin's, PF, B5*** | B Cell Lymphoma |
| Mac-2 | Macrophages/Histiocytes | No | Bouin's*** | Histiocytic Sarcoma |
| Lysozyme | Macrophages/Histiocytes/Myeloid Cells | Yes** | Formalin | Histiocytic Sarcoma |

PF = Paraformaldehyde

*Usually requires antigen retrieval or related methods for optimal immunoreactivity. For CD3, some antibodies which are useful in many species may be more optimal in mice.

**Depends on antibody used.

***Other quick-fixing preservatives can also be used with immediate embedding.

used in specialized applications, which include evaluation of immunoglobin deposits in the skin and in renal glomerular membranes, examination of the neuropeptides in nerves, and antigen studies using confocal microscopy.

Immunofluorescence is best used with frozen sections or fresh whole-cell preparations, because formalin-fixed tissues have a tendency to autofluoresce.    As
previously mentioned, the structural imperfections apparent in enzyme-labeled frozen sections have minimal effect on the presentation of immunofluorescence-labeled
tissues, as the unlabeled background tissue is minimally visible.

In the years since the introduction of immunofluorescence with fluorescein
isocyanate as its only fluorophore, many fluorescent labels have been developed.
Each requires a different filter set on the viewing microscope, because each excites
at a different wavelength. Fluorescein isothiocyanate (FITC) was introduced in
1958[13] as the new standard, and is still widely used today. It emits a bright-green
fluorescence at a wavelength of 495 nm. Rhodamine derivatives tetrarhodamine
isothiocyanate (TRITC) and Texas Red fluoresce red at 530 nm. Texas Red fades
less readily than TRITC.[14]    Phycoerythrin is also a rhodamine derivative, but it
fluoresces at the same wavelength as fluorescein, giving off a weak orange glow.
7-amino-4 methyl-coumarin-3-acetic acid (AMCA) is a blue fluorescence. It was
introduced as a fluorophore in 1986.[15]

## VI. INTERPRETING RESULTS

A brown or red signal for immunohistochemistry or bright green, red, or other
fluorochrome color for Immunofluorescence does not necessarily indicate a positive
reaction. Often apparent immunoreactivity may not be indicative of antigen expres-
sion/localization. The chromagen color may be indicative of nonspecific staining or
even a cross reaction of the antibody to an epitope in another antigen. Molecular
mimicry can be quite common. To determine if apparent specific staining is nonspecific, one must use proper controls. Some examples are included in Table 10.2.
Nonspecific binding of antibody or "background staining" can interfere with both
the interpretation of the reaction and its photographic quality for publication or
presentation.    There are several possible causes of background staining.

Some cells have very high levels of endogenous peroxidase and will yield false
positive reactions with the peroxidase enzyme. Mouse mast cells, for example,
particularly in the skin, often stain nonspecifically.[16] When typing cells in the skin,
a toluidine blue stain, commonly used to define mast cells, will differentiate mast
cells from those marked by the method used. Obvious endogenous peroxidase
activity may also be blocked by incubation of slides in a solution of hydrogen
peroxide and methanol, as mentioned previously under Basic Concepts. If nonspecific staining due to endogenous peroxidase continues to cause difficulty with interpretation of a reaction, one may consider changing the enzyme system used, which
should eliminate the problem.

Cells with a particularly strong positive charge, such as eosinophils, may bind
with negatively charged antibodies, causing nonspecific staining. Aldehyde fixation
may reduce the positive charge in tissues where this problem persists.

As a general rule, a hydrogen peroxide blocking step and the initial incubation
of tissues in normal serum will prevent most nonspecific staining from occurring.

Immunofluorescent background staining may be caused by autofluorescence as
well as nonspecific antibody reaction. Some tissues have been found to have a certain

degree of autofluorescence, such as alveolar macrophages, lipofuchsin, and elastic tissue, which may pose a problem in immunofluorescent reactions. Often the color or excitation wavelength of the intrinsic fluorescence differs from the fluorophore used in the reaction, so this problem may be easily negated. In some cases, a fluorescent counterstain such as 4-acetamido-4'- isothiocyanatostilbene-2,2' disulfonic acid (SITS), Fluoro Nissl Green, Evans Blue, or Acridine Orange may be helpful. Or a different fluorophore may be used. Evans Blue is commonly used with fluorescein, fluorescing red against fluorescein's green color.

When expression of an antigen is found in a tissue or tumor and one is uncertain of its significance or even if the result is real, supporting laboratory evidence is urgently needed. Western and northern blots can often provide important supporting evidence for gene expression in tissues and cells studied. If an antigen is found for the first time in a given tissue, northern blotting should show mRNA expression and Westerns would show protein expression. They generally correlate well with immunohistochemistry results.

## VII. MULTIPLE STAINING METHODS

Multiple staining allows the demonstration of more than one antibody within the same tissue. Using the avidin–biotin systems common today, there are two general methods of multiple staining. One may use the same enzyme system with different substrates to detect each antigen, or may choose to use a different enzyme system for each antigen localized. Adding antibodies sequentially and developing them fully before adding the next is the preferred method for multiple labeling with enzyme systems. In order to obtain optimal results, care must be taken with the order of labeling, as well as with the use of additional blocking steps for endogenous enzymes. Several controls may also be necessary to determine the success of each reaction.

Multiple staining with fluorescence-labeled primary antibodies is very simple, providing that a different fluorophore is used to label each antibody. These antibodies may be applied sequentially or simultaneously, and are viewed with separate filter systems.

## VIII. FURTHER INFORMATION

This chapter is designed to give a basic overview of immunohistochemical applications and techniques. For more detailed information, there are several books that may be helpful. *A Color Atlas of Dermatoimmunohistocytology*[17] has many fine color photographs and information on immunohistochemical and immunofluorescent techniques as specifically related to the study of skin. The book *Introduction to Immunocytochemistry*[18] contains everything you need to know to get started and become proficient at immunological histology. This is a great book for the beginner. *Techniques in Immunocytochemistry*[19-22] is a four-volume series containing a wealth of information. Chapters include discussions of both general techniques and specific applications. These books are more appropriate for a person with some background in immunohistochemistry to increase their knowledge. *Immunocytochemistry* [23] presents methodology and applications in a textbook format, and is easy to use as a reference and contains information for both the beginner and the experienced scientist. In addition, many of the vendors of immunohistochemical reagents offer booklets describing immunohistochemical methods in regard to the use of their products. These sometimes have very detailed diagrams and can often be very useful.

Several other important immunohistochemical references are listed below. [12,25-27]

## ACKNOWLEDGMENTS

This work was supported by a grant from the National Cancer Institute (CA34196).

## REFERENCES

1. Coons, A.H., Creech, H.J., and Jones, R.N., Immunological properties of an antibody containing a fluorescent group, *Proc. Soc. Exp. Biol. Med.,* 47, 200, 1941.
2. Coons, A.H. and Kaplan, M.H., Localization of antigen in tissue cells, *J. Exp. Med.,* 91, 1, 1950.
3. Coons, A.H., Leduc, E.H., and Connolly, J.M., Studies on antibody production, I:A method for the histochemical demonstration of specific antibody and its application to a study of the hyperimmune rabbit, *J. Exp. Med.,* 102, 49, 1955.
4. Nakane, P.K. and Pierce, G.B., Jr., Enzyme labeled antibodies: preparation and application for the localization of antigens, *J. Histochem. Cytochem.,* 14, 929, 1966.
5. Avrameas, S. and Uriel, J., Méthode de marquage d'antigène et d'anticorps avec des enzymes et son application en immunodiffusion, *C.R. Acad. Sci. Paris Sér. D.,* 262, 2543, 1966.
6. Mason, D.Y. and Sammons, R.E., Alkaline phosphatase and peroxidase for double immunoenzymatic labeling of cellular constituents, *J. Clin. Pathol.,* 31, 454, 1978.
7. Suffin, S.C., Muck, K.B., Young, J.C., Lewin, K., and Porter, D.D., Improvement of the glucose oxidase immunoenzyme technique, *Am. J. Clin. Pathol.,* 71, 492, 1979.
8. Bondi, A., Chieregatti, G., Eusebi, V., Fulcheri, E., and Bussolati, G., The use of β-galactosidase as a tracer in immunohistochemistry, *Histochemistry,* 76, 153, 1982.
9. Sternberger, L.A., Hardy, P.H., Jr., Cuculis, I.J., and Mayer, H.G., The unlabeled antibody–enzyme method of immunohistochemistry. Preparation and properties of soluble antigen–antibody complex (horseradish peroxidase–anti-horseradish peroxidase) and its use in identification of spirochetes, *J. Histochem. Cytochem.,* 18, 315, 1970.
10. Guesdon, J.L., Ternynck, T., and Avrameas, S., The use of avidin–biotin interaction in immunoenzymatic techniques, *J. Histochem. Cytochem.,* 27, 1131, 1979.
11. Hsu, S.M., Raine, L., and Fanger, H., Use of avidin–biotin–peroxidase complex (ABC) in immunoperoxidase techniques; a comparison between ABC and unlabeled antibody (PAP) procedures, *J. Histochem. Cytochem.,* 29, 577, 1981.
12. Kanai, K., Nunoya, T., Shibuya, K., Nakamura, T., and Tajima, M., Variations in effectiveness of antigen retrieval pretreatments for diagnostic immunohistochemistry, *Res. Vet. Sci.,* 64, 57, 1998.

13. Riggs, J.L., Seiwald, R.J., Burkhalter, J.H., Downs, C.M., and Metcalf, T., Isothio-cyanate compounds as fluorescent labeling agents for immune serum, *Am. J. Pathol.,* 34, 1081, 1958.
14. Titus, J. A., Haughland, R., Sharrows, S.O., and Segal, D.M., Texas Red, a hydro-philic, red emitting fluorophore for use with fluorescein in dual parameter flow microfluorimemetric and fluorescence microscopic studies, *J. Immunol. Meth.,* 50, 193, 1982.
15. Khalfan, H., Abuknesha, R., Rand-Weaver, M., Price, R.G., and Robinson, D., Ami-nomethyl coumarin acetic acid: a new fluorescent labelling agent for proteins, *Histochem. J.,* 18, 497, 1986.
16. Bussolati, G. and Gugliotta, P., Nonspecific staining of mast cells by avi-din–biotin–peroxidase complexes (ABC), *J. Histochem. Cytochem.,* 31, 1419, 1983.
17. Ueki, H. and Yaoita, H., *A Color Atlas of Dermatoimmunohistocytology,* Wolfe Medical Publications Ltd., London, 1989.
18. Polak, J.M. and Van Noorden, S., *Introduction to Immunocytochemistry,* 2nd edition, Springer-Verlag, New York, 1997.
19. Bullock, G.R. and Petrusz, P., *Techniques in Immunocytochemistry,* Vol. 1, Academic Press, London, 1982.
20. Bullock, G.R. and Petrusz, P., *Techniques in Immunocytochemistry,* Vol. 2, Academic Press, London, 1983.
21. Bullock, G.R. and Petrusz, P., *Techniques in Immunocytochemistry,* Vol. 3, Academic Press, London, 1986.
22. Bullock, G.R. and Petrusz, P., *Techniques in Immunocytochemistry,* Vol. 4, Academic Press, London, 1989.
23. Sternberger, L.A., *Immunocytochemistry,* 3rd edition, John Wiley & Sons, New York, 1986.
24. Jones, T.C., Ward, J.M., Mohr, U., and Hunt, R.D., *Hemopoietic System, Monographs on Pathology of Laboratory Animals,* Springer-Verlag, Berlin, New York, 1990.
25. Pileri, S.A., Roncador, G., Ceccarelli, C., Piccioli, M., Briskomatis, A., Sabattini, E., Ascani, S., Santini, D., Piccaluga, P.P., Leone, O., Damiani, S., Ercolessi, C., Sandri, F., Pieri, F., Leoncini, L., and Falini, B., Antigen retrieval techniques in immunohis-tochemistry: comparison of different methods, *J. Pathol.,* 183, 116, 1997.
26. Shetye, J.D., Scheynius, A., Mellstedt, H.T., and Biberfeld, P., Retrieval of leukocyte antigens in paraffin-embedded rat tissues, *J. Histochem. Cytochem.,* 44, 767, 1996.
27. Taylor, C. and Cote, R.J., *Immunomicroscopy: a Diagnostic Tool for the Surgical Pathologist,* 2nd edition, W. B. Saunders, Philadelphia, 1994.

## DETAILED PROTOCOLS

### IMMUNOHISTOCHEMICAL STAINING OF FROZEN TISSUE (SPECIFICALLY DEVELOPED FOR SKIN SECTIONS)

1. Air dry slides—5 min
2. Fix in acetone—10 min
3. Encircle sections with PAP pen (Kiyota International, Elk Grove Village, IL: cat# K-500)
4. Hydrate in PBS* (for any amount of time)

5.  Fix in Morpho-Save (Ventana Medical Supplies, Tucson, AZ: cat#250-010)—15 min
6.  Rinse with PBS
7.  Incubate with Peroxo-Block (Zymed, So. San Francisco, CA: cat#00-2015)—30 sec
8.  Rinse with PBS
9.  Apply a few drops of reagent A** and incubate—10 min
10. Rinse with PBS
11. Apply a few drops of reagent B** and incubate—10 min
12. Rinse with PBS
13. Block with TNB blocking buffer*** (0.5 gm blocking reagent/100 ml PBS)—30 min
14. Blot excess buffer from sections
15. Add biotinylated primary antibody[†] and incubate at room temperature for 1 hour or overnight at 4°C
16. Wash 3 times in TNT wash buffer (0.5 ml Tween 20[‡]/ 1L PBS)—5 min each
17. Incubate with HRP-SA*** (1:100 in TNB blocking buffer)—30 min
18. Wash 3 times in TNT washing buffer—5 min each
19. Incubate with BT working solution*** (1:50 in amplification diluent)—5 min
20. Wash three times in TNT wash buffer—5 min each
21. Incubate with Vectastain ABC-AP reagent (Vectastain ABC-AP Kit, Vector Laboratories: cat# AK-5000)—30 min
22. Wash 3 times in TNT washing buffer—5 min each
23. Incubate with a red chromagen designed for use with alkaline phosphatase. Two examples are AP-Red Substrate Kit (Zymed: cat# 00-2203) or HistoMark PtThalo Red Solution for Immunohistochemistry (Kirkegaard and Perry Laboratories, Gaithersburg, MD: cat# 71-00-01, 02, 04). Follow instructions with each chromagen.
24. Counterstain with hematoxylin
25. Mount in appropriate mountant. AP-Red Substrate requires an aqueous mountant; HistoMark Red may be mounted with either aqueous or xylene based mountant.

*From: Endogenous Avidin/Biotin Blocking Kit (Zymed: cat# 00-4303)
** Phosphate Buffered Saline (PBS) 10x (Life Technologies, Grand Island, NY: cat# 70013-032), diluted to 1x with distilled water.
*** From: TSA Indirect (NEN Life Science Products: cat# NEL700)
† Dilute primary antibody with TNB Blocking Buffer
‡ Tween 20 (Fisher Scientific, Pittsburgh, PA: cat# BP337-100)

**IMMUNOHISTOCHEMICAL STAINING OF PARAFFIN EMBEDDED, FIXED TISSUES**

1. Deparaffinize in two changes xylenes—5 min each
2. 1 change 1:1 xylenes/absolute ethanol—5 min
3. 1 change absolute ethanol—5 min
4. $H_2O_2$/Methanol (20 ml 30% $H_2O_2$ in 180 ml methanol) for blocking endogenous peroxidase—15 to 30 min
5. 1 change absolute ethanol—5 min
6. 1 change 95% ethanol—5 min
7. 1 change in 80% ethanol—5 min
8. 1 change 70% ethanol—5 min
9. Wash well in running filtered tap water
10. Encircle with PAP pen
11. Wash in PBS
12. Incubate tissues in normal (blocking) serum*—30 min
13. Blot excess serum from sections
14. Apply primary antibody per individual instructions — incubate in humidified chamber overnight at 4°C
15. Wash 3 times with TNT washing buffer (0.5 ml Tween 20/ 1L PBS)—5 min each
16. Apply biotinylated antibody* - 30 min at room temp. in humidified chamber
17. Wash 3 times with TNT washing buffer—5 min each
18. Apply ABC Reagent* - 45 min at room temp. in humidified chamber
19. Wash 3 times in TNT washing buffer—5 min each
20. Apply DAB (Vector Laboratories, cat# SK-4100) until proper color develops, from 30 seconds to a few minutes
21. Wash in running filtered tap water—3 min
22. Counterstain with Mayer's hematoxylin and coverslip per protocol

*From: Vectastain ABC Kit (Vector Laboratories, Burlingame, CA: (415) 697-3600)

**COUNTERSTAINING AND MOUNTING TISSUES (XYLENE-BASED MOUNTANT)**

1. Process slides and develop as described in immunohistology protocol.
2. Wash slides in water for 3 min.
3. Put slides in filtered hematoxylin for 2 to 3 min.
4. Wash in tap water for 5 min.
5. Remove slides from water and immerse in acid alcohol (1% HCl in 70% ethanol) twice (two 1-second immersions). The slides will go red.
6. Put slides into running tap water (it *must* be tap water, not distilled water). The slides will slowly turn from red to blue.
7. When slides look blue, put them into 70% ethanol for 3 min.

8. Put slides into 95% ethanol for 3 min.
9. Put slides into 100% ethanol for 3 min.
10. Put slides in xylene for 5 min, then change to clean xylene before cover-slipping.
11. Coverslip using synthetic mounting medium.

# 11 Radiolabeled cRNA and *In Situ* Hybridization

*Diane S. Keeney*

## CONTENTS

## I. INTRODUCTION AND SCOPE

The use of radiolabeled cRNA for *in situ* hybridization is most powerful to localize specific gene transcripts to a specific cell type(s), throughout the development of an organism. The data are three-dimensional. The cross-sectional area of the tissue section (*x, y* axes) provides spatial information. The third dimension (*z* axis) is depth in the case of serial sections of an organ or organism, or time in a developmental time course study. Mastering this technique can be daunting to the novice despite a plethora of published protocols. This review is intended to point out successful strategies and common pitfalls faced by the molecular biologist who has little or no experience with morphological methods, nuclear track emulsions, and photographic chemistry. Many of the methods have been adapted, with minor modifications, from the very successful protocol of Hogan et al.[1] Hence, this review is intended to supplement the radioactive *in situ* hybridization methods of these investigators[1] and to provide alternatives. The scope is limited to the use of radiolabeled cRNA to localize transcripts in paraffin-embedded tissue sections, although much of the discussion is applicable to frozen tissue sections.

## II. OVERVIEW

### A. WHY USE RADIOLABELED cRNA PROBES?

The overwhelming reasons to use radiolabeled cRNA for *in situ* hybridization are the excellent resolution and reproducibility of the method, and the high degree of specificity and sensitivity of these probes to detect relatively rare transcripts. Depending on the spatial distribution of transcripts within tissues, the sensitivity can approach that of reverse transcription-polymerase chain reaction, using total RNA as a template. Sequences having 90% identity are readily distinguished. For data presentation, the underlying cellular morphology is clearly visible using bright-field illumination. A major advantage is the use of dark-field illumination to discern global patterns of specific hybridization at low magnification, even when signal intensities are weak. Importantly, artifacts are readily distinguished from specific hybridization by the regular size, shape, and distribution of silver grains. A major disadvantage, however, is the length of time (2 to 4 weeks) required to obtain data from a single experiment. Another is that few investigators are trained in the use of nuclear track emulsions, autoradiography, and photographic chemistry, unlike the time when these were pioneering methods in cell cycle analysis. Inexperience can lead to problems, such as fogging and chemography effects in nuclear track emulsions, which are difficult to troubleshoot.

## B. PLANNING THE FIRST EXPERIMENT

Most technical problems arise because of inexperience with histological techniques (tissue manipulation, preservation, and analysis) and photographic chemistry. For abundant transcripts, success may come quickly. For less abundant or rare transcripts, persistence is crucial since 2 to 4 weeks are required to obtain results for every set of variables tested. It is important to use a positive control that is an abundant transcript demonstrated previously (by *in situ* hybridization) to be expressed in a specific cell type, at a specific stage of development. Predictions based on protein levels can waste valuable time. Best results are obtained when high and low levels of expression are present in different cells in the same tissue section, e.g., osteopontin in mouse embryos. It is difficult to distinguish signal from noise when most cells express the targeted transcript at similar levels, e.g., NADPH-cytochrome P450 reductase. It is preferable to start with a successful protocol from a colleague who can troubleshoot problems, rather than mixing protocols from different sources. Also, because it is difficult to write long protocols without errors, use common sense and double check calculations.

## C. GLOBAL PRECAUTIONS

- Physically separate reagents and materials for *in situ* hybridization from those used for other procedures, including other RNA methods.
- Clean glassware with a neutral detergent, rinse thoroughly, and bake at 240-260°C for $\geq$4 h.
- Whenever possible, use RNase-free, DNase-free, disposable plasticware, and sterilize solutions by filtration (0.22 μm) to avoid autoclaves, a potential source of RNase contamination.
- Treat water used for making solutions with diethyl pyrocarbonate, an RNAase inhibitor (DEPC-$H_2O$). Use very high-quality water at all steps.
- Be rigorous in cleanliness. Wear gloves and change them frequently. Cover work surfaces with clean bench paper. Keep work areas and equipment clean and dust-free.

These precautions are essential to simplify the process of troubleshooting failed experiments.

## III. THE METHODS

### A. TISSUE FIXATION

The first critical step in a successful *in situ* hybridization experiment is to obtain the very best tissue preservation possible. If there is any doubt what constitutes excellent preservation, consult an experienced morphologist. Variables in fixation protocols include fixative type, tissue type, tissue block size, fixative volume, penetration time, agitation schedule, temperature, etc. The fixation protocol affects not only the cutting properties of the embedded tissue but also the ability of the probe to find its target mRNA. Excessive cross-linking can hinder the penetration of cRNA into the tissue (weak signal), as well as the removal of excess, unbound probe (high background). Inadequate cross-linking leads to loss of cellular protein and nucleic acids (weak or no signal). For many tissues, success is obtained by immersion fixation in freshly prepared, 4% buffered paraformaldehyde, at 4°C for 2 to 24 h, depending on tissue type and block size.

Powdered paraformaldehyde (Sigma No. P-6148) dissolves rapidly at high pH but slowly at neutral pH. For rapid preparation, add NaOH to the water, dissolve the solute, and then add the phosphate-buffered saline (PBS). Determine empirically how much NaOH is required to achieve pH 7.4 after $20 \times$ PBS is added to $1 \times$ final concentration. For a starting point, add 4 g paraformaldehyde to a baked glass beaker containing 95 ml DEPC-H$_2$O and 50 µl 1.25 N NaOH. Heat the solution to 60°C while stirring on a hot plate, then turn off the heat. Do not overheat or boil. Continue stirring until the solute dissolves completely, $\approx$5 min at 60°C. Add 5 ml $20 \times$ PBS and chill on ice. Check the pH using indicator paper graded to 0.3 pH units (J.T. Baker Dual-tint special indicator paper), and adjust to pH 7.4 by adding dilute HCl, if necessary.

### B. MANUAL PROCESSING

Compared with frozen tissue sections, paraffin-embedded tissues afford greater morphological detail for cell type identification. For best results, optimize infiltration and dehydration protocols for different tissue types. Automated processing of samples submitted to a histopathology laboratory can yield poor results if one automated procedure is used for all specimen types. Hogan et al.[1] described a procedure for manual tissue processing. A modified version of that protocol is outlined below. For small or soft tissue specimens (e.g., mouse embryos, full-thickness mouse skin, endocrine glands), this procedure can be done in 2 to 3 days, and steps 12 through 14 can be replaced by an additional 100% paraffin incubation. For large specimens (e.g., a whole rat brain), one week might be required for proper infiltration. Steps 2 through 4 are skipped in many protocols, regardless of tissue type. Prepare ethanolic solutions in DEPC-H$_2$O unless indicated otherwise. Perform all steps from fixation to dehydration (100% ethanol) at 4°C, xylene infiltration at ambient temperature, and paraffin infiltration at 5 to 10°C above the melting point of the paraffin formulation. For example, Paraplast X-tra (Oxford Labware No. 8889-503002) melts at 50 to 54°C. Set the vacuum oven at 57 to 60°C. Vacuum is necessary to degas the paraffin and accelerate infiltration. Many variations of the following protocol are successful.

## 1. Flow Chart for Manual Dehydration and Paraffin Embedment

1.  Fix tissue in fresh 4% buffered paraformaldehyde, 2 to 24 h
2.  Wash in 1 × PBS to remove excess fixative, 60 min or overnight
3.  30% ethanol in PBS, 60 min
    *(PBS minimizes shrinkage artifacts due to hypotonicity)*
4.  50% ethanol, 60 min
5.  70% ethanol, 60 min
    *(Tissue can be stored in 70% ethanol at 4°C or -20°C. Prolonged storage in 100% ethanol causes tissue to become brittle and difficult to cut.)*
6.  90% ethanol, 60 min
7.  95% ethanol, 60 min
8.  100% ethanol, 60 min
9.  100% ethanol, 60 min
10.  Xylenes, 60 min
11.  Xylenes, 60 min
    *(Use separate xylenes for embedment, dewaxing, and dehydration.)*
12.  Xylenes:Paraffin (2:1), 60 min
13.  Xylenes:Paraffin (1:1), 60 min
14.  Xylenes:Paraffin (1:2), 60 min
15.  100% paraffin, 60 min
16.  100% paraffin, 60 min or overnight
17.  Embed in molds containing fresh 100% paraffin. If necessary, remelt to orient specimens. Cool the blocks to ambient temperature and store them in Ziploc® bags at 4°C to minimize dehydration.

## 2. Tips for Manual Dehydration and Paraffin Embedment

Tissue processing is easily done by hand using baked glass scintillation vials, a small vacuum oven, baked glass bottles for ethanolic solutions, histology grade xylenes, disposable Tri-Pour beakers, paraffin, and embedding molds and labels. Use the same glass vials for all steps; decant one solution and replace it with the next. Infiltration (fixative, ethanolic series, and xylenes) can be accelerated by gentle rocking. Rectangular Peel-A-Way embedding molds (22 × 30 × 30 mm; Polysciences No. 18646B) have an advantage over conventional flat tissue cassettes (Polysciences No. 21467) in that rectangular blocks can be placed in a microtome holder in different orientations. To make a holder for these molds, cut six rectangular holes in the lid of a Sigma shipping box (6.25 × 3.5 × 2.0 inches). Two of these boxes fit easily inside a small vacuum oven (e.g., Precision No. 31468). Prepare the specimen labels in advance, in pencil. Determine how the tissue will be sectioned, and then use sterile toothpicks to orient specimens appropriately in the molds. Tissue-Tek cold plates (Sakura Finetek No. 4650) are an inexpensive alternative to an embedding center.[1]

## C. PARAFFIN TISSUE SECTIONS

### 1. Strategies for Data Analysis

There are many advantages to covering the entire surface of the *in situ* hybridization slides with serially sectioned, paraffin "ribbons." First, if the same pattern of specific hybridization is observed in consecutive sections on the same slide, the signal is probably real. Second, serial sections make it likely to obtain publication-quality data in a single experiment. An air bubble or fiber may render one section unsuitable for photomicroscopy, but a clean adjacent section often exhibits the same expression pattern. Even the expression in a single cell layer can be detected in more than one 5 μm section. Third, serial sections add a third dimension ($z$ axis) to the data set. For example, an entire embryonic day 13.5 mouse can be serially sectioned and mounted on ≈16 microscope slides, and specific hybridization can be traced throughout the entire embryo. Finally, nonspecific hybridization can be controlled rigidly by using serial sections to compare results for sense and antisense cRNA on adjacent sections on the same slide. Immediately before the hybridization step, use a Pap pen (Polysciences No. 21841) to make a hydrophobic barrier between the left and right halves of microscope slides containing serially sectioned tissue. Apply hybridization mixtures containing sense cRNA to one compartment and antisense cRNA to the other compartment. Alternatively, two antisense probes can be applied to adjacent sections, on the same slide. It is imperative to apply each probe to alternating sides of the microscope slides since the emulsion is always thicker at one end, giving the appearance of more developed silver grains.

### 2. Tips for Cutting and Collecting Tissue Sections

Freshly embedded tissue blocks are soft and cut easily. Dehydrated blocks can be rehydrated in DEPC-$H_2O$ to improve their cutting characteristics. Cut 4 to 7 μm serial sections using disposable blades to minimize RNase contamination. Charged, precleaned microscope slides (e.g., Fisherbrand Superfrost Plus, No. 12-550-15) can be used without pretreatment. Simply rinse the slides with ethanol and wipe them dry to remove dust and fibers, immediately before use. Label slides on the frosted end with pencil or solvent-resistant marker (e.g., Securline Marker II/Superfrost, Polysciences No. 21947). Tissues will stick to the charged surface if they are dried completely on a slide warmer (42 to 45°C, overnight). Use a covered slide warmer since dust is a potential source of RNase activity.

If a flotation bath is used to collect sections, use a glass flotation tray that can be baked to destroy RNases, and do not cover the bath with aluminum foil. Metal oxides cause positive chemography in nuclear track emulsions. Alternatively, float tissue sections on DEPC-$H_2O$ applied directly to the slides using a disposable 60 cc syringe and 22 μm syringe filter (Costar No. 8110). Preheat the DEPC-$H_2O$-covered slides at least 5 min on the slide warmer to minimize trapped air bubbles beneath the tissue sections. As the sections expand, reposition them while removing excess water with a disposable 3 cc syringe and 27 gauge needle.

The following tools are useful for these procedures: sterile, cotton-tipped applicator sticks and xylenes to clean the microtome blade; razor blades to trim blocks; paintbrushes (#00 to #2), fine pointed forceps and a disposable (27 gauge) needle to manipulate the sections; ethanol in a squirt bottle and DEPC-$H_2O$ in a spray bottle to control temperature, humidity, and static electricity; sterile cotton gauze pads to clean tools; a bucket of ice and a beaker of DEPC-$H_2O$ to chill and hydrate tissue blocks. The black waxed surface on the inside of individually wrapped autoradiographic film (Kodak) is ideal for collecting ribbons and cutting them into desired lengths. For the fundamentals of tissue sectioning, consult Hogan et al.[1] and a basic histology textbook.

## D. *In Vitro* Transcription and Generation of Radiolabeled cRNA

### 1. Designing Templates for *In Vitro* Transcription

Ideally, choose a vector that has two RNA polymerase promoters so that the same plasmid can be used to generate antisense and sense (control for nonspecific hybridization) cRNA. Subclone inserts containing 100 to 500 bp of gene-specific sequence. Avoid heterologous probes. Choose a subcloning strategy that minimizes the length of the multiple cloning region between the RNA polymerase promoter and the gene-specific sequence. Select restriction enzymes to linearize the templates for *in vitro* transcription, avoiding those that generate 3' protruding ends (e.g., *Aat* II, *Apa* I, *Ban* II, *Bgl* I, *Bsp*1286 I, *BstX* I, *Cfo* I, *Hae* II, *HgiA* I, *Hha I, Kpn* I, *Pst* I, *Pvu* I, *Sac* I, *Sac* II, *Sfi* I, *Sph* I). If unavoidable, incubate the linearized template with T4 DNA polymerase or Klenow fragment of *E. coli* DNA polymerase to produce blunt ends, using conventional protocols.[2]

### 2. Purification of Linearized Templates

Prepare clean plasmid DNA (e.g., Qiagen maxi-prep kit or comparable product). Digest aliquots (10 to 30 μg) of plasmid DNA for 1 to 2 h with the appropriate restriction enzymes for synthesis of sense and antisense cRNA. Add more enzyme and incubate longer to ensure completion. Before stopping the reaction, resolve an aliquot (≈2 μg) of each digest on a thin agarose gel to confirm that the template is completely linearized and that it appears as a discrete band of the predicted size. An overloaded gel will reveal residual, uncut plasmid. Purify the templates using RNase-free reagents as described in detail[1,2] and outlined below. It is not necessary to gel-purify the templates, but it is very important to remove all traces of phenol and salts.

1. Extract the digests with an equal volume of Tris (pH 8) saturated phenol:chloroform:isoamyl alcohol 25:24:1.
2. Re-extract the aqueous phase with an equal volume of chloroform.
3. Precipitate the aqueous phase by adding 1/10 volume of 3 $M$ sodium acetate (pH 5.2) and two volumes of ice cold ethanol.
4. After centrifugation, wash the pellet with 70% ethanol, dry, and resuspend the DNA in 10 m$M$ Tris—1 m$M$ EDTA at a final concentration of 1 µg/µl. Estimate recovery at 65%.

## 3. The *In Vitro* Transcription Reaction

This is a critical step where a novice often has trouble. The protocol of Hogan et al.[1] generates radiolabeled cRNA of very high specific activity. A slightly modified version of that protocol is outlined below. A common misconception is that the cRNA must be full-length. While this is important, it should not be achieved by compromising specific activity. It is imperative to generate probes of the highest specific activity possible. The key is: ***do not add unlabeled UTP to the reaction.*** The concentration of ³⁵S UTP in the reaction is much lower than that of the other rNTPs. As ³⁵S UTP becomes incorporated into cRNA, if the concentration of this free nucleotide becomes rate-limiting, the radiolabeled cRNAs may terminate prematurely. Even if extension is incomplete, the specific activity of the cRNA will still be very high (in the absence of unlabeled UTP). Prepare two reactions: one for the sense template and one for the antisense template. Bring reagents to ambient temperature (except enzymes) and mix as indicated below:

| Reagent | Amount | Final Concentration |
|---|---|---|
| [³⁵S]UTP ≈125 µCi | 10 µl | ≈6.25 µ$M$ |
| (New England Nuclear, NEG-039H) | | |
| 5× transcription buffer (provided with enzyme) | 4 µl | |
| 200 m$M$ DTT | 1 µl | 10 mM |
| RNasin (Promega) | 1 µl | |
| Linear DNA template (1 µg) | 1 µl | |
| DEPC-H₂O | 1 µl | |
| 10 m$M$ @ rGTP, rATP, rCTP | 1 µl | 500 µ$M$ @ |
| RNA polymerase (T3, T7, or SP6) | 1 µl | |
| Total: | 20 µl | |

Mix the reactants by finger-tapping the tube and incubate at 37°C for 2 h. Concentrated preparations of polymerases can be used. Additional enzyme (1 µl) can be added at the midpoint. Purify the radiolabeled cRNA as outlined below and described in more detail by Hogan et al.[1] *Caution: Use high quality microfuge tubes to prevent radioactive contamination, and dispose of radioactive waste appropriately.*

1. Digest the DNA template by adding:
    1 µl   RNasin

              1 µl   tRNA (20 mg/ml)
               1 µl   RNase-free DNase (RQ1, Promega)
               Incubate at 37°C for 15 min.
   2. Extract the radiolabeled cRNA after adding:
               5 µl     200 m$M$ DTT
               5 µl     10% SDS
               55 µl    DEPC-$H_2O$
               10 µl    3 $M$ sodium acetate, pH 5.2
               100 µl   phenol:chloroform:isoamyl alcohol (25:24:1)
   3. Transfer ≈90 µl of the aqueous phase to a sterile microfuge tube. Add the
      following and precipitate the cRNA at least 60 min at -20°C:
               45 µl    5 M ammonium acetate
               400 µl   ethanol

After centrifugation, wash the pellet with 70% ethanol. Dissolve the dried cRNA in 100 µl of 200 m$M$ DTT. Count a 2 µl aliquot and dilute to 1 ×10[6] cpm/µl by adding deionized formamide to 50% final concentration. Add 200 m$M$ DTT to make up the remaining volume. Recount a 2 µl aliquot. It is important that the hybridization mixtures contain sense and antisense cRNAs at the same cpm/ml. Diluting the cRNA minimizes pipeting errors. Store the cRNA at -20°C until use. Follow the guidelines of Hogan et al.[1] to evaluate [35]S UTP incorporation.

For rapid analysis of the integrity of the radiolabeled cRNA, resolve 1 to 3 ×10[6] cpm of each probe on a very thin native agarose gel. Fix the gel in 10% trichloroacetic acid for 10 min. Dry the gel and expose it to autoradiographic film overnight. Although the molecular size of the [35]S cRNA cannot be estimated accurately by this method, the relative migration vs. a DNA ladder is informative when compared between experiments and DNA templates. A good probe will not always be a single discrete band; however, there should not be very high-molecular-size products or a generalized smear of radioactivity.

## 4. Choice of Isotope

The higher energy isotope [33]P UTP can be substituted for [35]S UTP to shorten exposure times.[3] However, owing to its shorter half-life, this advantage is lost if the [33]P isotope is not used as soon as possible after synthesis. By contrast, [35]S labeled cRNA can be used for up to one month with excellent results, if initially prepared from a fresh synthesis. The [33]P UTP isotope is more expensive, and, theoretically, its higher energy (longer pathlength) yields less resolution.

## 5. Limited Hydrolysis of Radiolabeled cRNA

Hogan et al.[1] described methods for limited hydrolysis to reduce the size of the radiolabeled cRNA to 100 to 200 bp. To calculate the duration of hydrolysis, the size (kb) of the radiolabeled cRNA must be known. This step can be tricky. The length of the radiolabeled cRNA should be, but is not necessarily the same as that predicted from the linearized cDNA template. When a probe is being used for the first time, size the transcripts on denaturing polyacrylamide gels, before and after hydrolysis. Failure to do so may result in the loss of radiolabeled cRNA at this step, caused by overestimation of hydrolysis time and complete hydrolysis.

This step is more problematic for long transcripts, which have a greater probability of premature termination. In practice, excellent results can be obtained without hydrolysis even with cDNA inserts >1 kb. Clearly, limited hydrolysis should be done if satisfactory results are not obtained without hydrolysis. Ideally, when using a probe for the first time, compare results ± limited hydrolysis. The simplest approach, however, is to construct templates to contain short, gene-specific inserts that do not require limited hydrolysis and that generate full-length cRNA under the reaction conditions. Full-length transcripts provide more stringent control for nonspecific hybridization because, if designed properly, the sense and antisense cRNAs will encode opposite strands of the same sequence.

## E. Prehybridization

Prehybridization procedures aim to permeabilize the tissue sections and to block nonspecific binding sites. These procedures facilitate the diffusion of cRNA into the tissue and binding to its targeted mRNAs, as well as the diffusion of excess, unbound cRNA out of the tissue in the posthybridization wash. Prehybridization protocols are optimized for specific fixation methods. High background can result when different fixatives are used in conjunction with a single prehybridization protocol. Prehybridization treatments are done conveniently in baked glass Wheaton staining dishes (250 ml capacity; with glass lids and trays; Wheaton No. 900200). The glass trays (10 slots) hold 20 slides paired back-to-back or placed in a zigzag fashion. *Caution: Wheaton dishes are sensitive to rapid temperature changes. They can be baked safely at 260°C with incremental temperature changes (e.g., 90°C, 180°, 260°C). It is important, that they be undisturbed until cooled to ambient temperature.*

The flow chart below contains minor modifications of the prehybridization protocol of Hogan et al.[1] One significant difference is the use of pronase E (Protease type XIV; Sigma No. P-5147; 112 µg/ml final concentration) in place of proteinase K.[1] The number at each step refers to a single Wheaton dish containing the specified solution (17 dishes in this scheme). "Refresh" means that fresh solution should be added *before* reusing the same dish. Ethanolic solutions are indicated simply by the % ethanol in DEPC-$H_2O$. The PBS is 1× final concentration in DEPC-$H_2O$. "TEA/AA" indicates 0.1 $M$ triethanolamine buffer (Sigma No. T-1377) to which 0.25% acetic anhydride (Sigma No. A-6404) is added immediately before use. If two racks (40 slides) are processed in one experiment, most of the solutions (except for some of the lower % ethanolic solutions) should be added fresh for the second rack. Discard the 100% ethanol after deparaffinization to prevent carryover of residual xylenes and paraffin.

| | | |
|---|---|---|
| 1. | Xylenes I | 10 min |
| 2. | Xylenes II | 10 min |
| 3. | 100% I | 5 min |

|  | | |
|---|---|---|
| 4. 100% II | 5 min |
| 5. 95% | 4 min |
| 6. 85% | 4 min |
| 7. 75% | 4 min |
| 8. 65% | 4 min |
| 9. 50% | 4 min |
| 10. 30% | 2 min |
| 11. PBS I | 5 min |
| 12. 4% buffered paraformaldehyde | 20 min |
| 13. PBS II | 5 min |
| 13. PBS II *refresh* | 5 min |
| 14. Pronase E | 7 min |
| 13. PBS II | 5 min |
| 12. 4% buffered paraformaldehyde | 5 min |
| 15. DEPC-H$_2$0 | few seconds |
| 16. TEA/AA | 10 min |
| 17. TEA/AA | 10 min |
| 11. PBS I | 5 min |
| 10. 30% | 2 min |
| 9. 50% | 4 min |
| 8. 65% | 4 min |
| 7. 75% | 5 min |
| 6. 85% | 5 min |
| 5. 95% | 5 min |
| 4. 100% II *refresh* | 5 min |
| 3. 100% I   *refresh* | 10 min |

Air dry the slides 30 to 60 min before hybridization. Cover them loosely (inverted "V") with clean aluminum foil to keep out dust.

## F. HYBRIDIZATION

### 1. Cover Glasses

Hybridization should be done using cover glasses or another type of coverslip.  At 50°C, evaporation is noticeable by 24 to 36 h even with cover glasses. Evaporation is pronounced without cover glasses and introduces unacceptable variability in the experimental results. Several alternatives to cover glasses are available, including commercial RNase-free plastic covers and parafilm strips.[1] Cover glasses must be siliconized to reduce surface tension and trapping of air bubbles. If a PAP pen is used to divide the slides into compartments, siliconized cover glasses are easily cut to size by scoring with a diamond pencil.

Siliconize the cover glasses (24 × 60 mm; Corning No. 2935-246) well in advance. The following protocol yields excellent results without pretreatment to inhibit RNases, provided that the cover glasses are handled only with clean forceps and clean, gloved hands. First, use fine, pointed forceps to immerse the cover glasses individually in a baked glass Coplin staining jar (Wheaton No. 900520) containing Sigmacote (Sigma No. SL-2). Sigmacote should be used near a fumehood. Dry the cover glasses in a vertical position (e.g., against a test tube rack covered with clean aluminum foil). Re-dip the cover glasses in 100% ethanol. Dry them thoroughly in a vertical position (avoid dust) and repack into the original box. Sigmacote reserved for this purpose can be reused if filtered between uses to remove dust and fibers (e.g., baked glass funnel and Whatman paper).

## 2. The Hybridization Mixture

Prepare the hybridization mixture from frozen stocks, just before use. Prepare more than actually needed—a generous estimate is 120 to 150 μl per microscope slide. The protocol below was adapted from Hogan et al.[1]

|  | **for 1 ml hybridization mixture** |
| --- | --- |
| 10× Salts | 100 μl |
| Deionized formamide | 400 μl |
| 50% dextran sulfate | 200 μl |
| tRNA (20 mg/ml) | 10 μl |
| DTT (1 $M$) | 8 μl |
| DEPC-$H_2O$ | 82 μl |
| Diluted probe | 200 μl |

Add radiolabeled cRNA at a final concentration of 25 $\times 10^6$ cpm/ml of hybridization mixture. Good results can be obtained within the range 20 to 40 $\times 10^6$ cpm/ml, but this should be determined experimentally. After calculating the total cpm needed, calculate the volume of each radiolabeled cRNA (previously diluted to $\approx 10^6$ cpm/μl) that must be added. This volume should be much less than the 200 μl allotted for diluted probe above. The difference between these two volumes should be added as a mixture of deionized formamide and 200 m$M$ DTT (50:50).

Heat the hybridization mixture containing radiolabeled cRNA at 80 to 100°C, for 2 to 3 min, before use. Vortex vigorously and place on ice. Apply the mixture with a pipet tip (0 to 200 μl tip) held nearly parallel to the microscope slide. Use the broad, lateral surface of the pipet tip to spread the mixture evenly, as it is pushed out of the tip. Wet all of the tissue sections with the hybridization mixture (*no air bubbles*) before applying a cover glass.

A simple, humidified hybridization chamber can be made using plastic Tupperware containers.[1] Alternatively, use a plastic desiccator box (Bel-Art No. H42053-0001) placed inside a forced-air oven (e.g., Gallenkamp Plus II). Fill the desiccator tray with DEPC-$H_2O$ instead of desiccant and preheat the unit (50 to 55°C). Apply hybridization mixture and cover glasses to the slides and place them directly on the removable plastic shelves in the desiccator box. Two racks of 20 slides each are easily accommodated. Equilibrate the contents of the box to the preset oven temperature (50 to 55°C) for 10 to 15 min before sealing the door for the overnight incubation ($\approx 18$ h).  This same oven can be used to bake the cleaned glassware (260°C) at the end of the experiment.

## G. Posthybridization

The flow chart below contains minor modifications of the posthybridization protocol of Hogan et al.[1] These wash conditions are relatively stringent, generating low background levels with homologous cRNA probes. To modify the stringency, consult Hogan et al.[1] Prepare the wash buffers in baked glass bottles, from premade stock solutions. Preheat these solutions in waterbaths, as specified. The wash incubations are done conveniently in baked glass Wheaton staining dishes, placed inside a small, forced-air hybridization oven (e.g., Techne Hybridiser HB-1E). Move the slides quickly from the hybridization chamber to the first wash solution. The DTT concentrations (steps 1 and 2) used in different protocols vary from 10 to 100 m$M$. Because this reagent is expensive (unless purchased in bulk; Research Organics), lower concentrations are desirable as long as the background does not increase. Since DTT is labile, add it to the prewarmed wash solutions shortly before use. In step 6, background grain development increases significantly at salt concentrations >0.3× SSC, which explains why heterologous probes are generally not useful.

1. Incubate in 5× SSC—10 m$M$ DTT at 55°C, for 15 min to remove cover glasses.
2. Incubate in 50% formamide--2X SSC—100 m$M$ DTT at 55°C, for 20 min. *Subsequent steps do not require DEPC-H$_2$O.*
3. Incubate in TEN buffer at 37°C, for 10 min.
4. Incubate in TEN buffer containing RNase A (20 µg/ml final concentration) at 37°C, for 30 min.
5. Incubate in TEN buffer at 37°C, for 10 min.
6. Incubate in 0.2× SSC—1 m$M$ DTT at 60°C for 10 min, three times.
7. Incubate in 0.2× SSC at ambient temperature for 10 min.
8. Dehydrate in ethanolic solutions (30%, 60%, 80%, 90%, and 95%) containing 0.3 $M$ ammonium acetate final concentration, 4 min each.
9. Dehydrate in 100% ethanol for 10 min, two times.
10. Air dry for 1 h (dust-free) before coating with nuclear track emulsion.

## H. Nuclear Track Emulsions and Autoradiography

The remaining steps are often problematic for investigators who have never worked with nuclear track emulsions. The first decision is which emulsion to use. The properties of Ilford and Kodak nuclear track emulsions are quite different, and protocols for their use are not interchangeable. If you select an Ilford emulsion, consult Hogan et al.[1] The following protocols are for Kodak NTB emulsions. Select NTB-2 (Eastman Kodak No. 165 4433) for [35]S cRNA probes. This is probably also the best choice for [33]P probes. Consult Rogers[4] for theoretical and practical aspects of nuclear track emulsions. Kodak's technical services are outstanding (1-800-243-2555). Ask to speak to an expert on a topic of interest. Unfortunately, Kodak has discontinued many of its booklets and publications on the use of nuclear track emulsions and autoradiography.[5] These are excellent references if copies can be found.

## 1. Working with Nuclear Track Emulsions

Work in the darkroom in complete darkness to learn all of the sources of static sparks, which emit light and expose the emulsion. The exceptions are when a safelamp is needed to measure the emulsion or read a timer during emulsion processing. Use only the manufacturer's recommended lamp and filter combination (e.g., Kodak safelight lamp Model B, No. 141-2212; Kodak safelight filter No. 2, No. 152-1525) and work at 4 ft distance from the source. Relative humidity of approximately 50% minimizes static spark potential. If the darkroom is equipped with a sink, humidify the atmosphere by constant, slow-running, hot water. The following discussion assumes that the darkroom is not equipped with a revolving or double door system to allow repeated entry in complete darkness. If experiments are done frequently, stock a laboratory cart reserved for darkroom use with all of the equipment necessary for coating slides and processing emulsion.

### a. Liquefy the emulsion

First, prepare and set up the materials and equipment required to liquefy the emulsion. Preheat a water bath to 43 to 45°C in the darkroom (e.g., Lab-Line Lo-Boy tissue flotation bath with a Pyrex Petri dish insert, Corning No. 3140-190, 190 × 100 mm). The emulsion temperature will be slightly lower than that of the bath (optimally 42°C). This temperature should be constant within and between experiments for reproducible, uniformly thin emulsion films. Use NTB-2 emulsion diluted 1:1 (v:v) in water containing 2% glycerol. A total volume of 12 ml or 18 ml of liquefied emulsion is sufficient to coat 20 or 40 slides, respectively, using a Dip Miser slide coating cup (Electron Microscopy Sciences No. 70510). Note that Kodak advises against adding anything to the emulsion because of the potential for additives or contaminants to cause chemography (see troubleshooting section and Reference 4). In practice, neat NTB-2 emulsion produces films that are too thick for acceptable photomicroscopy at higher magnifications.

Prepare a measuring device for the emulsion by circumscribing the 12 ml line (for 20 slides) of a graduated, 50 ml centrifuge tube with a thin strip of colored laboratory tape. When read like a graduated cylinder, this line will serve as a guide to hold the tube level in the dark and to adjust the final volume of diluted emulsion. Add 6 ml of 2% glycerol in very pure, deionized $H_2O$ to this tube and take it to the darkroom in a tube rack. Work in the darkest corner of the darkroom (if all light leaks cannot be sealed) with the safelamp pointed away from the emulsion. In the dark, use a porcelain spoon or nonmetal spatula to cut the emulsion gel into small pieces. Kodak emulsions are not shredded like Ilford emulsions. Transfer pieces of emulsion into the centrifuge tube containing 6 ml of 2% glycerol until the water level is raised, by displacement, to the 12 ml line. Cap the tube *tightly,* and very gently turn it on its side. Wrap the tube in three layers of heavy-duty aluminum foil and float it in the preheated (43 to 45°C) water bath for 45 to 60 min. It is extremely important to avoid agitation that generates air bubbles. The foil makes it possible to leave the darkroom without exposing the emulsion.

## b. Coating the hybridized slides with emulsion

Order the hybridized slides into a black slide box (VWR Scientific No. 48444-004). Prepare packets of indicating Drierite by folding a small mound of desiccant inside one half of a 4 × 4-inch gauze pad. Tape the desiccant packet to a microscope slide and insert one packet into each of the black slide boxes to be used for drying/exposing the emulsion-coated slides. Prepare a chilled flat surface which will be used to gel the film of emulsion on the hybridized slides. For this purpose, cover a discarded glass DNA sequencing plate with clean, heavy-duty aluminum foil. Rubber bands can be used to divide the plate into three vertical lanes (Caution: these can generate static sparks). Return to the darkroom with these materials when the emulsion is melted. Place the glass plate on top of a half-inch bed of ice inside a large Nalgene pan (Nalgene No. 6900-0020). Level the plate and place it comfortably close to the water bath containing the liquefied emulsion. Place a large, wet sponge next to the Nalgene pan. This will be used to wipe emulsion from the back of the slides. Finally, place the Dip Miser cup in its metal holder in the heated water bath.

In darkness, pour the liquid emulsion into the Dip Miser cup. Coat three clean (blank) microscope slides with emulsion and examine them under the safelamp for air bubbles. When none are detected, dip the hybridized slides individually in a slow, smooth rhythm. To achieve uniform thickness, count 3 or 5 seconds for each downstroke, soak, and upstroke. Slower rates of withdrawal yield thinner films. Touch the slide to the lip of the cup to catch the drops, and immediately turn the slide horizontal, allowing the emulsion to flow uniformly across the slide. Wipe the back of the slide on the wet sponge. Carefully place the coated slides in a predetermined order on the chilled plate. It is important to align the frosted edge of each slide with the edge of the glass plate (or bands demarcating internal lanes). While more uniform emulsions can be obtained by drying the emulsion at ambient temperature (slides horizontal), often this is not practical because of the time restraints of common-use darkrooms, limitations of darkroom design, and potential for light leaks and accidental exposure.

When all of the slides are coated, pick them up in the same order in which they were coated and pack them into black plastic slide boxes containing desiccant. When all of the slides are in the boxes with the lids closed, seal the boxes with one-inch-wide laboratory tape *(Caution: source of static sparks)*. The boxes should be airtight and light-tight. Cover each box with two layers of heavy duty aluminum foil and store at ambient temperature until the emulsion is completely dry. For 20 slides, dry overnight or 24 h with the addition of fresh desiccant (in complete darkness) during this interval. Expose the slides at 4°C, typically for 10 to 15 days. Shorter exposures are required for abundant transcripts, such as collagen type I and type II, or epidermal scatterbrains. Coat a few "test slides" to judge the optimal exposure time. Discard the unused portion of liquefied emulsion.

## 2. Processing the Emulsion

### a. Photographic chemistry

If you selected an Ilford emulsion, consult Hogan et al.[1] The following protocol is for Kodak NTB-2 emulsion. Prepare the photographic chemistry using high-quality, deionized H$_2$O, following the manufacturer's instructions. Prepare the whole packet of developer and fixer, because the chemical granules are not uniform in size or composition. Do not use rapid fixer. Kodak recommends against an acid stop bath, since this can cause microbubbles in the emulsion. Bring the slides to ambient temperature. Set up four Wheaton staining dishes in an iced water bath prepared in the same Nalgene pan that was used for coating the slides. Prechill to 15°C the developer (Kodak D19 diluted 1:1 (v:v) in H$_2$O), stop bath (H$_2$O), and fixer (2 dishes). When *all* solutions are exactly 15°C, turn off the light and transfer the emulsion-coated slides to a glass Wheaton tray (20-slide capacity). Remove foil and tape from all the slide boxes before opening the first box to avoid exposing the emulsion to static sparks. Carry out steps 1 through 3 below in darkness. Do not agitate at any step. Steps 4 through 16 can be done in the light, back in the laboratory. Discard these solutions after one use.

1. D-19 (diluted 1:1, 15°C) (Kodak No. 146 4593)          2 min
2. Deionized H$_2$O (15°C)                                30 sec
3. Kodak Fixer (15°C) (Kodak No. 197 1738)               2 × 5 min
4. Wash the slides in slowly running, deionized H$_2$O or in     30 min
   several changes of a large volume of deionized H$_2$O
5. Stain in 0.005% Toluidine blue in 10 m$M$ borax, prepared   10 to12 min
   fresh from stock solutions for each use
6. Deionized H$_2$O                                        brief
7. 30% Ethanol                                            2 min
8. 50% Ethanol                                            4 min
9. 65% Ethanol                                            4 min
10. 75% Ethanol                                           5 min
11. 85% Ethanol                                           5 min
12. 95% Ethanol                                           5 min
13. 100% Ethanol                                          5 min
14. 100% Ethanol                                          10 min
15. Xylenes I                                             10 min
16. Xylenes II                                            10 min

These dehydration times can be shortened, but artifacts will occur if dehydration is not complete. Remove slides individually from the last xylenes. Immediately apply Permount (Fisher Scientific No. SP15-100) or comparable permanent mounting medium and a No. 1$\frac{1}{2}$ grade cover glass (Corning No. 2940-245), without allowing the xylenes to evaporate. Dry the Permount completely or at least overnight before taking the slide near a microscope objective.

### b. Choice of stain

Toluidine blue O (C.I. 52040) is a basic dye that stains predominantly nuclei at dilute concentrations. Cytoplasm stains lightly, which is ideal for visualizing developed silver grains representing cytoplasmic mRNAs. Alkaline buffers favor tissue binding and retention of this basic dye during destaining and dehydration in ethanolic solutions. Toluidine blue is also metachromatic, which is advantageous in identifying certain cell types (e.g., mast cells). For each tissue type, optimize conditions to retain the dye in the tissue but not in the emulsion. Overstaining requires extensive destaining to remove the dye from the emulsion. Some protocols deal with this problem by using unbuffered Toluidine blue.[1] Other stains that reveal more morphological detail are useful, but hematoxylin and eosin, for example, can cause quenching in dark-field illumination. Several trials may be necessary to identify optimal parameters for publication-quality data.

## c. Tips on photographic development

The development protocol largely determines the background level of silver grain development in the emulsion. Instructions in the NTB-2 package insert are for processing undiluted emulsion films. In practice, if the emulsion was diluted (thinner films), the developer is diluted. Hogan et al.[1] recommend much more dilute developer to suppress background in Ilford emulsions. This effectively suppresses noise, but it also suppresses specific signals and can lead to loss of very weak signals. This is advantageous if specific hybridization is intense. Ideally, one would like to significantly increase the signal:noise ratio, but this is not realistically achieved at this step, since the developer acts on both exposed and unexposed silver grains. It just works faster on exposed silver grains.

For reproducibility, prepare the photographic chemistry fresh for each experiment. Store the solutions in tightly capped, brown glass bottles with no air overlay to minimize oxidation. When all of the slides have been processed, discard unused portions. While D19 developer is active for weeks or months, its potency changes with time, causing unacceptable variation in the signal intensities between *in situ* hybridization experiments. Regarding the strength or potency of Kodak fixer, Kodak recommends the following method to estimate the optimal fixation time. With the lights on, immerse a dried, emulsion-coated slide in the fixer—**do not develop it first**. Multiply by a factor of two the minutes required for the emulsion to dissolve completely, i.e., from milky white to transparent.

## d. Troubleshooting emulsion artifacts

Many problems will be averted by consulting Kodak materials[5] and Reference 4 before the first experiment. Because nuclear track emulsions are very sensitive to chemical contaminants, especially metals and photochemicals, Kodak recommends three controls for every experiment: blank, charged slides (no tissue), charged slides with unlabeled tissues (not hybridized), and charged slides with hybridized (radioactive) tissue sections.

If artifacts are unavoidable (e.g., fogging), determine whether the problem arose during the steps between (a) emulsion dilution/liquefaction and emulsion processing or (b) between tissue preparation and the posthybridization wash. The first alternative contains the fewest steps and is easiest to test. If Kodak's recommended controls were not done, start with the following experiment. Coat cleaned, charged (or subbed) microscope slides without tissue sections in *undiluted* emulsion, as recommended by the manufacturer. Dry the slides and develop them photographically. If the artifact disappears, the artifact observed earlier very likely resulted from a contaminant introduced during emulsion dilution or processing.

Try the obvious: purchase new chemicals, make new stock solutions, try different sources or batches of water and photochemicals. If the artifact cannot be reproduced with charged slides without tissue sections, repeat the experiment ± tissue sections (not hybridized but deparaffinized, hydrated, and air dried before coating with emulsion). Finally, if the artifact still has not been reproduced, a contaminant may have been introduced between tissue preparation and the posthybridization wash (alternative b, above). In this case, repeat the entire *in situ* hybridization experiment using all of the recommended controls.

If the artifact does not disappear in the experiment with *undiluted* emulsion above, the emulsion itself must be suspected, along with other variables. Has the emulsion expired? Was it exposed to high temperatures, frozen, physically stressed, shocked, or refrigerated with radioisotopes? All of these should be avoided. Routinely use Plexiglas shields during storage of the emulsion stock and coated slides. Possibly, the emulsion was damaged during handling, storage, or transit before it was received. This is why Kodak recommends coating test slides for every new batch of emulsion before use in experiments.

## 3. Data Presentation

The first critical step in obtaining publication-quality images of *in situ* hybridization data is to understand the proper use of the light microscope and the principles of Köhler illumination. The Kodak publication P-2, *Photography Through the Microscope*, is an excellent guide. It can be obtained from Silver Pixel Press (1-800-368-6257) or from a local photo dealer. Carl Zeiss Inc. offers outstanding technical service and training in the use of the light microscope, photomicroscopy, and image capture, when a microscope is purchased through a Zeiss representative. If purchasing a microscope for the analysis of radioactive *in situ* hybridization data, the following Zeiss Plan Neofluar objectives, or comparable products from another vendor, are recommended: 2.5×/0.075, 5×/0.15, 10×/0.30, 20×/0.50 PH2, and 40×/0.75 D. The top lens of the substage condenser must be flipped out when using objectives <10×. For 5 to 7 μm paraffin sections, 40× is the highest power objective that is reasonably needed. However, for thinner paraffin sections, and if the microscope will be used for other applications, choose a Plan Apochromat 63×/1.40 oil.

A high-intensity, fiber-optic illuminator is indispensable for dark-field illumination (MVI; Micro Video Instruments, Inc.). This light source utilizes a stage adaptor to transmit light through the ends of the microscope slide. A dark-field stop in the substage condenser is still very useful; however, the MVI fiber-optic illuminator is essential for viewing large specimens. If very large specimens will be photographed routinely (e.g., whole mouse embryos), use the Plan Neofluar 1.25×/0.04 in conjunction with the MVI fiber-optic illuminator and the Zeiss dark-field illuminator (Zeiss No. 445214; for use at 1.25× to 20×) that fits in the condenser carrier, in place of the substage condenser. This combination effectively addresses two problems commonly associated with imaging a specimen using wide diameter (<10×) objective lenses: low contrast images and difficulty in obtaining very sharp focus.

The Zeiss dark-field illuminator makes it possible to capture high-contrast, low-magnification images. Its main disadvantage is that it must be removed, and the substage condenser replaced, to capture bright-field images of the same specimen. A focusing telescope (Zeiss No. 522012) is also useful in obtaining sharper images at <10×.

To photograph the same field in bright-field and dark-field illumination, use Kodak Ektachrome 64T (35 mm slide) film with two neutral density filters (e.g., 3% and 12%). This tungsten film gives true color without color filters; that is, what you see through the microscope is what you see on the developed slide. For color print film, Fujicolor Reala (ASA 100; daylight film) gives reasonably true color with a blue filter. Neutral density filters are still needed with Reala to accommodate the same field in bright-field and dark-field illumination. For color prints, it is sometimes possible to find a local photo developer franchise that specializes in one-hour processing that will match the color on the prints to that of a sample print. This is feasible because once the *in situ* hybridization method is working well, the results are very reproducible, even the color in the final data analysis.

## IV. TIME COURSE OF AN EXPERIMENT

The synthesis date of the isotope is the logical anchor in organizing an *in situ* hybridization experiment. Typically, this is the first week of each month for [$^{35}$S]. During this week, prepare the tissue sections and radiolabeled cRNAs. Plan two consecutive days to complete steps from prehybridization to coating the slides with emulsion. On the morning of the first day, do the prehybridization treatments. Apply the hybridization mixture in the afternoon, and incubate the slides overnight. On the morning of the second day, preheat the posthybridization wash buffers. Wash the slides in the afternoon. While the slides are drying, set up the darkroom equipment and equilibrate the water bath. Liquefy the emulsion and coat the slides the same night. The next day, when the emulsion has dried completely, put the slides at 4°C to expose. In 7 to 10 days, make fresh photographic chemistry and develop a few test slides to evaluate the exposure time. Using the same chemistry, develop, stain, and dehydrate the remaining slides. Apply cover glasses and dry the permanent mounting medium overnight. When the slides are dry, analyze the data. Discard unused photographic chemistry.

## V. REAGENTS AND RECIPES

### DIETHYL PYROCARBONATE-TREATED WATER (DEPC-H$_2$O)

Add 1 ml diethyl pyrocarbonate (Sigma No. D5758) to a 1 L bottle containing very clean, deionized H$_2$O (0.1% final concentration). Shake vigorously to dissolve. Autoclave the next day to remove all traces of DEPC.

## 20× Phosphate buffered saline, pH 7.4 (PBS)

For one liter, dissolve the following in DEPC-$H_2O$. Adjust the pH to 7.4 with NaOH [prepare NaOH in DEPC-$H_2O$]. Filter sterilize. Use at 1× final concentration.

|  | amount | 20×concentration |
|---|---|---|
| NaCl (FW=58.44) | 163 g | 2.8 $M$ |
| KCl (FW=74.55) | 4 g | 54 m$M$ |
| $KH_2PO_4$ (monobasic FW=136.1) | 4 g | 29 m$M$ |
| $Na_2HPO_4$ (dibasic, heptahydrate FW=268.1) | 43.43 g | 162 m$M$ |

## 4% buffered paraformaldehyde

Make fresh each day. For 100 ml, dissolve 4 g paraformaldehyde (Sigma No. P-6148) in 95 ml DEPC-$H_2O$ containing 50 µl of 1.25 N NaOH. Heat to 60°C while stirring—**do not overheat**. When completely dissolved, add 5 ml 20× PBS and chill on ice. Adjust the pH to 7.4, if necessary.

## 3 $M$ sodium acetate, pH 5.2 (NaOAc)

For 50 ml, dissolve 12.3 g NaOAc (FW=82.03, anhydrous) in 40 ml DEPC-$H_2O$. Adjust the pH with glacial acetic acid. Adjust the volume to 50 ml. Filter sterilize.

## 5 $M$ ammonium acetate ($NH_4OAc$)

For 100 ml, dissolve 38.5 g $NH_4OAc$ (FW=77.08) in DEPC-$H_2O$. **Filter sterilize—Do not autoclave.** Store at 4°C.

## 1 $M$ dithiothreitol (DTT)

Dissolve 3.08 g DTT (FW=154.24) in 20 ml 0.01 $M$ NaOAc, pH 5.2 [66 µl 3 $M$ NaOAc in 20 ml DEPC-$H_2O$]. **Filter sterilize—do not autoclave.** Store aliquots at -20°C. Dilute 1:5 with DEPC-$H_2O$ for 200 m$M$ working stock.

## 10% sodium dodecyl sulfate (SDS)

Dissolve 5 g SDS in 50 ml DEPC-$H_2O$.

## Yeast transfer RNA (tRNA)

Dissolve 55 mg tRNA (BRL Gibco No. 15401-029) in 2.75 ml DEPC-$H_2O$ for 20 mg/ml stock solution. Follow the manufacturer's instructions. Store aliquots at -20°C. Use this preparation of tRNA without purification.

## Pronase E (Protease type XIV from *Streptomyces griseus;* Sigma No. P-5147)

Dissolve 1 g pronase E in 25 ml sterile, DEPC-H$_2$O for a stock concentration of 40 mg/ml. Predigest at 37°C, for 4 to 5 h, according to the manufacturer's instructions. Store 700 µl aliquots at -20°C. One aliquot is sufficient for 20 slides when diluted into 250 ml protease buffer (one Wheaton dish; 112 µg/ml final concentration). *Note:* Sigma now offers a molecular biology grade product (No. P-6911).

### Protease buffer

Make fresh for each use, from stock solutions. For 250 ml, add:

|  | amount | final concentration |
|---|---|---|
| 1 *M* Tris base pH 7.4 | 12.5 ml | 50 m*M* |
| 0.5 *M* EDTA, pH 8 | 2.5 ml | 5 m*M* |
| DEPC-H$_2$O | 235 ml | |

## 1 *M* Tris, pH 7.4

For 100 ml, dissolve 12.1 g Tris base (FW=121.1) in deionized H$_2$O. (Residual DEPC in DEPC-treated water can react with Tris buffers.) Adjust the pH to 7.4 with HCl. Filter sterilize.

### Deionized formamide, for the hybridization mixture

Use high-quality, molecular biology grade, deionized formamide. Store aliquots at -20°C (e.g., in baked glass scintillation vials).

### 50% Dextran sulfate, for the hybridization mixture

Dissolve 15 g dextran sulfate (MW 500,000; Sigma No. D-8906) in 30 ml DEPC-H$_2$O. Aliquot into sterile 2 ml microfuge tubes. Store at -20°C. Use wide-bore pipet tips. Beware of pipeting errors due to viscosity.

### 5 *M* sodium chloride, for the hybridization mixture

For 50 ml, dissolve 14.6 g NaCl (FW=58.44) in DEPC-H$_2$O. Filter sterilize.

### 10× salts, for the hybridization mixture

For 50 ml, mix the following:

|  | amount | 10× concentration |
|---|---|---|
| 5 *M* NaCl | 30 ml | 3 *M* |
| 1 *M* Tris, pH 7.4 | 5 ml | 100 m*M* |
| 0.5 *M* EDTA, pH 8 | 5 ml | 50 m*M* |
| Na$_2$HPO$_4$ (dibasic, heptahydrate, FW=268.1) | 1.34 g | 100 m*M* |
| Ficoll 400 (MW 400,000, Sigma No. F-2637) | 0.1 g | 0.2% |
| Polyvinylpyrrolidone (Sigma No. P-5288) | 0.1 g | 0.2% |
| BSA fraction V | 0.1 g | 0.2% |

Add DEPC-H$_2$O to 50 ml, dissolve, and filter sterilize. Aliquot into sterile microfuge tubes and store at -20°C. Thaw once and discard unused portion. Vortex until there is no precipitate. Do not heat to thaw. This will denature the BSA.

## 0.5 *M* DISODIUM ETHYLENEDIAMINETETRAACETATE, pH 8 (EDTA)

For 500 ml, dissolve 93.05 g EDTA (dihydrate, FW=372.25) in 400 ml DEPC-H$_2$O. Stir vigorously while adding ≈10 g NaOH pellets, until the pH is 8.0. Adjust the volume to 500 ml. Filter sterilize.

## 20× SSC, pH 7

For 1 L, dissolve the following in DEPC-H$_2$O. Adjust the pH to 7. Filter sterilize.

|  | amount | 20× concentration |
| --- | --- | --- |
| Sodium chloride (FW=58.44) | 175.3 g | 3 *M* |
| Sodium citrate (dihydrate, FW=294.1) | 88.2 g | 0.3 *M* |

## 0.1 *M* TRIETHANOLAMINE (TEA BUFFER)

Make fresh for each use. For 500 ml buffer, dissolve 6.64 ml (or 7.46 g) of triethanolamine in DEPC-H$_2$O. This liquid (FW=149.2, Sigma No. T-1377) has a density=1.124 g/ml. Add acetic anhydride (Sigma No. A-6404) to this buffer immediately before use, to a final concentration of 0.25%. Add 625 μl acetic anhydride to 250 ml TEA buffer in a baked glass bottle. Shake vigorously to dissolve. This is sufficient for 20 slides in a Wheaton dish.

## TEN BUFFER, FOR RNASE DIGESTION

Prepare fresh from stock solutions for each use. For 250 ml, add the following to deionized water:

|  | amount | final concentration |
| --- | --- | --- |
| 1 *M* Tris pH 7.4 | 2.5 ml | 10 m*M* |
| 0.5 *M* EDTA | 2.5 ml | 5 m*M* |
| 5 *M* NaCl | 25 ml | 500 m*M* |

## 5 *M* NaCl FOR TEN BUFFER

For 500 ml, dissolve 146.1 g NaCl (FW=58.44) in deionized H$_2$O. Sterilize.

## RNASE A

Dissolve 50 mg RNase A (Sigma No. R-6513) in 0.5 ml sterile, deionized H$_2$O for a working concentration of 100 mg/ml. Pipet 50 μl aliquots into sterile, 0.5 ml microfuge tubes, and store at -20°C. Avoid prolonged storage, which can lead to lyophilization of this small volume. One aliquot is sufficient for 20 slides when diluted into 250 ml TEN buffer (one Wheaton dish; 20 μg/ml final concentration). It is no longer necessary to boil RNase A preparations. Follow the manufacturer's instructions. The pH optimum for RNase A activity is 7.0 to 7.5.

## ETHANOLIC SOLUTIONS CONTAINING 0.3 $M$ NH$_4$OAc

For 1 liter, mix the following:

| % ethanol | ml ethanol | ml 5 $M$ NH$_4$OAc | ml deionized H$_2$O |
|-----------|-----------|--------------------|---------------------|
| 30% | 300 | 60 | 640 |
| 60% | 600 | 60 | 340 |
| 80% | 800 | 60 | 140 |
| 90% | 900 | 60 | 40 |
| 95% | 940 | 60 | 0 |

## 0.005% TOLUIDINE BLUE O—10 m$M$ BORAX

1. Make a 1% (w:v) stock solution of toluidine blue in deionized H$_2$O. Leave a stir bar in the bottle to mix the solution before taking an aliquot.
2. Make a 2% (w:v) stock solution of borax (FW=381) in deionized H$_2$O.
3. Prepare 10 m$M$ borax by diluting 38 ml of 2% borax in deionized H$_2$O to a total volume of 200 ml. Add 1 ml of 1% toluidine blue. This is sufficient for 20 slides in a Wheaton staining dish.

## REFERENCES

1. Hogan, B., Beddington, R., Costantini, F., and Lacy, E., *Manipulating the Mouse Embryo, A Laboratory Manual, Second Edition*, Cold Spring Harbor Press, Plainview, 1994 (Section H).
2. Sambrook, J., Fritsch, E.F., and Maniatis, T., *Molecular Cloning, A Laboratory Manual, Second Edition,* Cold Spring Harbor Press, Plainview, 1989 (Section 17.23).
3. McLaughlin, S.K. and Margolskee, R.F., [33]P is preferable to [35]S for labeling probes used in *in situ* hybridization, *Biotechniques*, 15, 506, 1993.
4. Rogers, A.W., *Techniques of Autoradiography, Third Edition,* Elsevier/North-Holland Biomedical Press, Amsterdam, 1979.
5. Kodak Publication P-64, *Kodak materials for light microscope autoradiography;* Publication M6-110, *Photographic procedures for light and electron microscope autoradiography, a selected bibliography of books and papers*; and the unpublished booklet entitled *Microautoradiography: Autoradiography at the light microscope level.*

# 12 Use of Nonradiolabeled Probes for *In Situ* Hybridization

*Brian J. Limberg and Charles C. Bascom*

## CONTENTS

## I. INTRODUCTION

Development of nonradiolabeled probes for use in *in situ* hybridization (ISH) occurred during the 1980s.[1] Biotin-UTP or digoxigenin-UTP are now commonly used in the synthesis of nonradiolabeled nucleic acid probes.[2-4] The advantages for employing nonradiolabeled probes for *in situ* hybridization studies are the improved stability of the nonradiolabeled probe, greater cellular resolution for the hybridization signal, an improved signal-to-noise ratio, and the decreased time to observe the hybridization signal. With these advantages, nonradiolabeled probes have been more widely employed in recent years. For example, digoxigenin (DIG)-labeled probes have been used to localize expression of keratin 17,[5] type I hair keratin genes,[6] involucrin,[7] fibroblast growth factor receptor and ligand genes,[8] and mouse Notch[9] to skin and hair follicles.

Thus, it also seems appropriate to review methods for ISH using a nonradiolabeled probe; in particular, the use of DIG-labeled riboprobes for *in situ* hybridization will be covered. A standard detection kit is readily available from Boehringer-Mannheim (the Genius™ Nucleic Acid Labeling and Detection Kit, Indianapolis, IN). Boehringer-Mannheim also has developed a *Non-radioactive In Situ Hybridization Application Manual* (Second Edition) to review applications for DIG-labeled DNA- or RNA-labeled probes.

0-8493-1905-6/00/$0.00+$.50
© 2000 by CRC Press LLC

## II. TISSUE PREPARATION

Regardless of the signal detection method employed, precautions mentioned in the previous chapter should still be exercised when collecting tissues for nonradiolabeled ISH studies. Again, one should take steps to eliminate possible contamination by RNases by baking glassware, autoclaving solutions, and using RNase-free solutions (i.e., diethylpyrocarbonate (DEPC)-treated water, RNase-free DNase, etc.). The steps in preparing the target tissue for *in situ* hybridization—i.e., fixation and sectioning—have already been thoroughly discussed. However, a majority of the ISH work performed in our laboratory has been with murine and human skin samples, so a few comments will be made with respect to tissue preparation. Immersion fixation with fresh, ice-cold, buffered 4% paraformaldehyde works well for skin samples. Incubation times in 4% paraformaldehyde solution can vary, but usually 6 h to overnight incubations at 4°C work well (see Table 12.1). Following fixation of the skin samples, paraffin-embedding, performed as described in the previous chapter with radiolabeled probes for ISH, also works well with skin samples. Cryosectioning of skin sections for ISH studies can also be performed, but the tissues are prepared in a different way. Immersion fixation in fresh 4% paraformaldehyde is still performed; however, the tissues are then placed in a cold, 30% sucrose solution in phosphate-buffered saline overnight, which improves the cutting of the tissue and minimizes tissue damage caused by ice crystals.[10] The tissues are then placed in an embedding matrix, such as M-1 embedding matrix (Shandon-Lipshaw, Pittsburgh, PA; Cat. No. 1310). Because tissue morphology is better with paraffin-embedded tissues, our laboratories use paraformaldehyde-fixed, paraffin-embedded skin sections for the majority of the ISH studies (see Table 12.1). Finally, SuperFrost Plus slides (Fisher Scientific, Pittsburgh, PA; Cat. No. 12-550-15) work well with skin sections (usually around 5 to 7 µm).

The steps most crucial in preparing the tissue sections for hybridization are the proteolysis step, which permeabilizes the tissue so the labeled cRNA can gain access to its cellular target, and steps taken to reduce background signals (see Table 12.1). Most protocols employ the use of proteinase K to permeabilize the target tissue. The amount of proteinase K used should be empirically determined with each tissue; however, the concentration of proteinase K commonly used ranges from 1 to 50 µg/ml.[11] In addition, the length of incubation in the proteinase K solution can vary from 5 to 30 minutes, and incubations are done either at room temperature or at 37°C. To inactivate the proteinase K, some protocols also include a brief incubation in a Tris-glycine solution.

## III. SYNTHESIS OF THE NONRADIOLABELED RIBOPROBE

The *in vitro* transcription reaction incorporating DIG-UTP into cRNA is performed essentially as described with either $^{35}$S-UTP or $^{33}$P-UTP. However, the concentration of the DIG-labeled riboprobe is determined by spotting serial dilutions of the purified DIG-labeled cRNA onto nitrocellulose, along with a standard DIG-labeled control RNA (Boehringer Mannheim, Indianapolis, IN; Cat. No. 1 585 746). This provides

**TABLE 12.1**
*In Situ* Hybridization Protocol Using DIG-Labeled Probes

| Process | Procedure | Method | Suggested Conditions |
|---|---|---|---|
| TISSUE | Tissue fixation | 4% PFA/PBS, pH 7.2 | 4°C, 6 hr to overnight |
| PROCESSING | | 70% ethanol | 4°C, 1 hr to overnight |
| | Embedding | Paraffin-embedding | Paraplast X-tra |
| SLIDES | | | FisherBrand SuperFrost Plus slides |
| TISSUE SECTION | Tissue sectioning, 5-7 μm | | |
| PREPARATION | Deparaffinization | Heat sections | 65°C, 1 hr |
| | Rehydration | Xylene | Room temperature, 2 × 10 min |
| | | 100% ethanol | Room temperature, 2 × 5 min |
| | | 95% ethanol | Room temperature, 2 × 5 min |
| | | 80% ethanol | Room temperature, 1 × 5 min |
| | | 70% ethanol | Room temperature, 1 × 5 min |
| | | PBS | Room temperature, 2 × 5 min |
| | Tissue permeabilization | Proteinase K in PBS | 37°C, 5-30 min, 1-50 μg/ml |
| | | *Tris-HCl, pH 7.0 (132 mM)-glycine (100 mM) | Room temperature, 1 × 5 min |
| | | PBS | Room temperature, 2 × 5 min |
| | Post-fixation | 4% PFA/PBS, pH 7.0 | Room temperature, 10 min |
| | | PBS | Room temperature, 2 × 5 min |
| | Acetylation | Acetic anhydride (0.25%) in 0.1 M triethanolamine, pH 8.0 | Room temperature, 1 × 10 min |
| | | PBS | Room temperature, 2 × 5 min |
| | Dehydration | 70% ethanol | Room temperature, 1 × 1 min |

**TABLE 12.1**
*In Situ* Hybridization Protocol Using DIG-Labeled Probes (continued)

| Process | Procedure | Method | Suggested Conditions |
|---|---|---|---|
| | | 95% ethanol | Room temperature, 1 × 1 min |
| | | 100% ethanol | Room temperature, 1 × 1 min |
| HYBRIDIZATION | Prehybridization | Formamide (50%) | Room temperature, 1 hr |
| | | 5× SSC | |
| | | 5× Denhardts | |
| | | tRNA (250 μg/ml) | |
| | | Carrier DNA (500 μg/ml) | |
| | | **thioUMP (28.6 μg/ml) | |
| | | BSA (100 μg/ml) | |
| | | **DTT (5 m*M*) | |
| | Hybridization | Probe in hybridization buffer | Overnight, $T_m$ –15-25°C, 50-60°C |
| | | **(prehybridization without thioUMP) | DIG-labeled probe, 5-200 ng/ml |
| | Remove coverslips | 2× SSC | Room temperature, 5 min |
| | Posthybridization | 0.2 × SSC, Formamide (50%) | 2 × 30 min, 60-70°C |
| | washes | 0.2× SSC | Room temperature, 1 × 10 min |

* Optional
**Optional; used mainly with $^{35}$S-radiolabeled probes

only a semiquantitative assessment of the concentration of the DIG-labeled ribo-probe, so methods to improve quantitation of the DIG-labeled cRNA have been developed.[12]

## IV. PREHYBRIDIZATION AND HYBRIDIZATION CONSIDERATIONS

Various steps are taken to reduce nonspecific binding of a probe to the slide and target tissue (see Table 12.1). These steps involve an incubation in 0.1 $M$ triethano-lamine containing 0.25% acetic anhydride, in order to reduce nonspecific binding of either radiolabeled or DIG-labeled probes to the slide or to the tissues. Reagents also commonly added to the prehybridization buffer to reduce background binding to tissue sections include carrier DNA, tRNA, Denhardt's, and bovine serum albumin (BSA). For the nonradiolabeled probes, the concentration of the DIG-labeled probe added to the hybridization buffer can also affect the signal-to-noise ratio. The concentration of DIG-labeled probes ranges from 5 to 200 ng/ml.

There are several factors to consider when establishing the hybridization and posthybridization conditions, including pH, temperature, salt concentration, the nature of the probe, and the composition of the hybridization and washing solutions.[11] The optimum temperature for hybridization is usually 15 to 25°C below the melting temperature ($T_m$) of the nucleic acid hybrids, and a formula for calculating the $T_m$ when performing RNA:RNA hybridization has been determined.

$T_m$ = 79.8 + 18.5(log[Na]) + 58.4(G–C fraction) + 11.8(G–C fraction)$^2$ – (820/# base pairs duplexed) – 0.35(% formamide) – % mismatch

Because RNA-RNA hybrids are more stable than DNA-RNA or DNA-DNA hybrids, more stringent hybridization and washing conditions can be performed with DIG-labeled riboprobes.

## V. DETECTION OF THE NONRADIOLABELED PROBE

Immunological detection of the DIG-labeled probe is performed using the Genius™ system developed by Boehringer-Mannheim (Indianapolis, IN) (see Table 12.2). Following posthybridization washes, the sections are first equilibrated in Buffer 1 [0.1 $M$ maleic acid/0.15 $M$ NaCl (pH 7.5)], blocked for one hour in Buffer 2 (1% blocking powder in Buffer 1), and then incubated with an anti-DIG antibody (ranging from a 1:1000 to a 1:4000 dilution) in Buffer 2 for one hour at room temperature. The sections are washed three times in Buffer 1 for 20 minutes at room temperature. The sections are then equilibrated in 0.1 $M$ Tris/0.1 $M$ NaCl/0.05 $M$ MgCl$_2$ (pH 9.5) (Buffer 3), and the color reaction is performed by adding 67 μl of 4-nitroblue tetrazolium chloride (NBT; Boehringer-Mannheim, Indianapolis, IN; Cat. No. 1 383 213) and 70 μl 5-bromo-4-chloro-3-indolyl-phosphate (BCIP) (Boehringer Man-nheim, Indianapolis, IN; 1 383 221) per 20 ml of Buffer 3. The reaction is stopped by placing the sections in a 0.1 $M$ Tris/1 m$M$ EDTA (pH 8.0) solution. To avoid

**TABLE 12.2**
**Color Detection Protocol for Standard ISH or Tyramide Signal Amplification**

| Process | Procedure | Method | Suggested Conditions |
|---|---|---|---|
| COLOR | Blocking | Boehringer-Mannheim kit | Room temperature, 1 hr |
| DETECTION FOR | | Blocking reagent (1% in Buffer 1) | |
| DIG-LABELED | | *goat serum (1%) | |
| PROBES | | *Levamisole (1.5 m$M$) | |
| (DIG-ISH) | Antibody | Blocking reagent (1% in Buffer 1) with anti-digoxigenin antibody | Room temperature, 1 hr |
| | | (1:1000 to 1:4000 dilution) | |
| | Wash | Buffer 1 | Room temperature, 3 × 20 min |
| | Rinse | Buffer 3 | Room temperature, 1 × 5 min |
| | Color development | Buffer 3 with NBT (67.5 µl/20 ml) | In dark, room temperature, |
| | | BCIP (70 µl/20 ml) | 1 hr to overnight |
| | | *Levamisole (1.5m$M$) | |
| TYRAMIDE | Blocking | Boehringer-Mannheim kit | Room temperature, 1 hr |
| SIGNAL | | Blocking reagent (1% in Buffer 1) | |
| AMPLIFICATION | | *goat serum (1%) | |
| COLOR | | *Levamisole (1.5 m$M$) | |
| DETECTION | Primary Antibody | Blocking reagent (1% in Buffer 1) with biotin-labeled, anti-digoxigenin antibody (1:500) | Room temperature, 1 hr |
| | Wash | TNB blocking buffer | Room temperature, 3 × 5 min |

| | | |
|---|---|---|
| Secondary conjugate | Streptavidin-horseradish peroxidase (1:100) in TNB blocking buffer | Room temperature, 30 min |
| Wash | TNT Buffer | Room temperature, 3 × 5 min |
| Tyramide precipitation | Biotinylated-tyramide (1:25) in 1× Amplification Diluent | Room temperature, 15 min |
| Wash | TNT Buffer | Room temperature, 3 × 5 min |
| Block | TNB Blocking Buffer | Room temperature, 30 min |
| Tertiary conjugate | Streptavidin-alkaline phosphatase (1:100) in TNB Blocking Buffer | Room temperature, 30 min |
| Wash | TNT Buffer | Room temperature, 3 × 5 min |
| Color development | Buffer 3 with NBT (67.5 µl/20 ml) | In dark, room temperature, |
| | BCIP (70 µl/20 ml) | 1 hr |
| | *Levamisole (1.5mM) | |

* Optional

distortion of the colorimetric signal, the slides are mounted using an aqueous-based mounting media such as Crystal Mount (BioMeda Corp., Foster City, CA; Cat. No. M02), placed in a 65°C oven until the media has hardened, then the slides are cover-slipped using Permount (Fisher Scientific, Pittsburgh, PA; Cat. No. SP15-100).

One of the disadvantages commonly discussed with DIG-labeled cRNA probes is sensitivity. While it is generally considered that radiolabeled probes provide greater sensitivity over DIG-labeled probes, Komminoth et al.[1] have claimed that DIG-labeled probes can provide sensitivity comparable to $^{35}$S-labeled probes. Nevertheless, a recent method has been developed to improve the sensitivity with DIG-labeled riboprobe. This is known as tyramide signal amplification (TSA),[13] and is based on methods developed to enhance signals when performing immunoassays or immunohistochemistry (also referred to as catalyzed reporter deposition, or CARD).[14, 15]

Tyramide signal amplification is based on the fact that tyramide will precipitate in the presence of horseradish peroxidase. Thus, the ISH signal is amplified in two ways: via detection of biotin-labeled anti-DIG antibodies (Sigma Chemical Co., St. Louis, MO.) and biotinylated tyramide. The protocol for detecting the DIG-labeled probe by TSA is described in the Renaissance TSA-INDIRECT Kit (NEN Life Science Products, Boston, MA; Cat. No. NEL730) (also see Table 12.2). Briefly, following posthybridization washes, the sections are incubated in TNB Buffer (0.1 $M$ Tris-HCl, pH 7.5; 0.1 $M$ NaCl; 1% blocking powder [provided with Renaissance TSA-INDIRECT KIT]) for one hour. A biotin-labeled, anti-DIG antibody (1:500 dilution) in TNB buffer is added to the sections for one hour. The sections are washed, and then incubated in the TNB buffer containing a streptavidin–horseradish peroxidase conjugate (1:100 dilution). After washing the sections in TNT buffer (0.1 $M$ Tris-HCl, pH 7.5; 0.1 $M$ NaCl; 0.05% Tween-20), biotinylated tyramide (1:25) in 1x Amplification Diluent is added to the sections for 15 minutes. The sections are washed in TNT buffer, again placed in TNB buffer for 30 minutes, and then a streptavidin-alkaline phosphatase conjugate (1:100) in TNB buffer is added to the tissue sections for 30 minutes. Following another three washes in TNT buffer, the color reaction is performed by adding 67 µl NBT and 70 µl BCIP per 20 ml of Buffer 3. The reaction is stopped by placing the sections in a 0.1 $M$ Tris/1 m$M$ EDTA (pH 8.0) solution. At this point, the slides may be counterstained using hematoxylin for 1 to 2 minutes, rinsed with water, then allowed to dry. Once the slides have dried they can be mounted using the methods described previously.

If background problems develop with this protocol, additional reagents can be added to the buffers. To reduce background binding of anti-DIG antibodies to tissue, one can add inactivated goat serum (1%) to blocking buffer; or to block endogenous alkaline phosphatase activity in tissues, one can also add levamisole (1.5 m$M$) to the blocking solution. Also, technical representatives for both Boehringer-Mannheim and NEN Life Science Products are available to address any technical issues involving their kits.

## VI. TIME COURSE OF AN EXPERIMENT

Three aspects need to be considered when planning ISH with nonradiolabeled probes. First, synthesis of the nonradiolabeled probe can be performed anytime, because the probes are stable for up to one year when stored at -80°C. Setting up the *in vitro* transcription reaction and purifying the cRNA probe requires at least 2 to 3 h. It also takes about 2 to 3 h to determine the concentration of the Nonradio-labeled probe. Second, it usually takes one-half day to process the tissue sections and to perform the prehybridization and hybridization steps. Third, depending on the development method used (standard DIG-ISH vs. TSA), it takes one-half to one day to process the tissues, perform the color detection, and counterstain the tissue sections.

## REFERENCES

1. Komminoth, P., Merk, F.B., Leav, I., Wolfe, H.J., and Roth, J., Comparison of ³⁵S- and digoxigenin-labeled RNA and oligonucleotide probes for *in situ* hybridization. Expression of mRNA of the seminal vesicle secretion protein II and androgen receptor genes in the rat prostate, *Histochemistry*, 98, 217, 1992.
2. Bloch, B., Biotinylated probes for *in situ* hybridization histochemistry. Use for mRNA detection, *J. Histochem. Cytochem.*, 41, 1751, 1993.
3. Schaeren-Wiemers, N. and Gerfin-Moser, A., A single protocol to detect transcripts of various types and expression levels in neural tissue and cultured cells: *In situ* hybridization using digoxigenin-labelled cRNA probes, Histochemistry, 100, 431, 1993.
4. Fisher, C., Angus, B., and Rees, J., *In situ* hybridization using digoxigenin-labelled probes in human skin, *Br. J. Dermatol.*, 125, 516, 1991.
5. Panteleyev, A.A., Paus, R., Wanner, R., Nürnberg, W., Eichmuller, S., Thiel, R., Zhang, J., Henz, B.M., and Rosenbauch, T., Keratin 17 gene expression during the murine hair cycle, *J. Invest. Dermatol.*, 108, 324, 1997.
6. Winter, H., Siry, P., Tobiasch, E., and Schweizer, J., Sequence and expression of murine type I hair keratins mHa2 and mHa3, *Exp. Cell Res.*, 212, 190, 1994.
7. de Viragh, P.A., Huber, M., and Hohl, D., Involucrin mRNA is more abundant in human hair follicles than in normal epidermis, *J. Invest. Dermatol.*, 103, 815, 1994.
8. Rosenquist, T.A. and Martin, G., Fibroblast growth factor signalling in the hair growth cycle: Expression of the fibroblast growth factor receptor and ligand genes in the murine hair follicle, *Develop. Dynamics*, 205, 379, 1996.
9. Kopan, R. and Weintraub, H., Mouse notch: Expression in hair follicles correlates with cell fate determination, *J. Cell Biol.*, 121, 631, 1993.
10. Tokuyasu, K.T., A technique for ultracrytomy of cell suspensions and tissues, *J. Cell Biol.*, 57, 551, 1973.
11. Jin, L. and Lloyd, R.V., *In situ* hybridization: Methods and applications, *J. Clinical Lab. Anal.*, 11, 2, 1997.
12. Didenko, V.V. and Hornsby, P.J., A quantitative luminescence assay for nonradioactive nucleic acid probes, *J. Histochem. Cytochem.*, 44, 657, 1996.

13. Kerstens, H.M.J., Poddighe, P.J., and Hanselaar, A.G.J.M., A novel *in situ* hybridization signal amplification method based on the deposition of biotinylated tyramine, *J. Histochem. Cytochem.*, 43, 347, 1995.

14. Bobrow, M.N., Harris, T.D., Shaughnessy, K.J., and Litt, G.J., Catalyzed reporter deposition, a novel method of signal amplification. Application to immunoassays, *J. Immunol. Meth.*, 125, 279, 1989.

15. Bobrow, M.N., Shaughnessy, K.J., and Litt, G.J., Catalyzed reporter deposition, a novel method of signal amplification. II. Application to membrane immunoassays, *J. Immunol. Meth.*, 137, 103, 1991.

# 13 Repositories of Mouse Mutations and Inbred, Congenic, and Recombinant Inbred Strains

*Muriel T. Davisson and John J. Sharp*

## CONTENTS

## I. INTRODUCTION

It is widely recognized that genetically defined mice provide (1) reproducible, experimental systems for understanding normal development and function, (2) model systems for analyzing the defects in comparable human disorders, and (3) model systems for preclinical testing of therapeutic agents. Model systems permit studies that are inappropriate or impossible in human beings. The mouse, in particular, provides a good model system because of the metabolic and internal anatomical similarities between mice and human beings, the availability of controlled genetic backgrounds, short life span, large litters, and short generation time. Because of their ease of maintenance and small size, mice are relatively economical to maintain. Their cells and tissues are readily accessible for studies of gene dosage effects on embryogenesis and organogenesis. Mouse embryogenesis is well studied and described. More is known about the genetics of the mouse than any other experimental mammal, and specific regions of the mouse and human genomes are highly conserved.[1-3]

The Jackson Laboratory (TJL) maintains an extensive collection of genetically defined mice and provides them to the scientific community. Over 2300 stocks are available as breeding mice, frozen embryos, or DNA samples, and over 1200 spontaneous and induced mutations are maintained at TJL. The types of strains maintained include inbred strains, strains and stocks carrying induced or spontaneous mutations, stocks carrying chromosomal aberrations, recombinant inbred strains, and congenic inbred strains (strains with selected alleles maintained on specific genetic backgrounds). Each of these genetically defined types of mice has specific uses. This chapter describes the different types of genetically defined strains of mice and their availability from TJL.

## II. HISTORY OF GENETIC AND ANIMAL RESOURCES AT THE JACKSON LABORATORY

The Jackson Laboratory has served as a central repository to identify, develop, protect, preserve, and distribute mutant mice and inbred strains for over 50 years. The first inbred strain (DBA; dilute brown nonagouti) was developed by Clarence Cook Little before he founded the laboratory in 1929. Jackson Laboratory scientists shared their inbred strains with other investigators until the level of distribution justified a separate Production and Distribution Department, which was established in the mid-1950s by then Director, Earl L. Green. The first individual research colonies of spontaneous mutants at TJL were established in the 1930s by George D. Snell and Elizabeth S. Russell. In the late 1940s, Margaret Dickie assembled many of these mutants into a single colony and established the Mouse Mutant Stocks Center (MMSC). A second colony consisting primarily of neurological mutants was formed by Earl L. Green in the 1960s. When Eva M. Eicher assumed responsibility for this colony in 1971 and added non-neurological mutants and chromosomal aberrations, it became the Mouse Mutant Gene Resource (MMGR). The MMSC and MMGR were consolidated into the present Mouse Mutant Resource (MMR) in 1983, under the direction of Muriel T. Davisson.[4] The Induced Mutant Resource (IMR) was established in September 1992, with John J. Sharp as supervisor.[5,6] Scientific staff members also maintain individual colonies of mutant mice for research purposes. Altogether, TJL holds more than 1200 mutant gene bearing stocks.

George D. Snell conceived of and began developing histocompatibility congenic strains during the 1940s,[7] for which he received the Nobel Prize in 1980. Donald W. Bailey conceived of and began developing recombinant inbred strains in 1981.[8] The need to protect genetically defined, often unique strains of mice from accidental loss resulted in a program to cryopreserve mouse embryos in 1976. The banking of frozen embryos was begun by Donald W. Bailey and Larry E. Mobraaten in 1978 and continues today under the direction of Dr. Mobraaten.[9]

New strains are continually being added to TJL's large collection by new strain development (see MMR below) and by importation of strains from outside TJL (see

IMR below). All mice that are imported into TJL are processed through an established importation program, designed to free incoming mouse strains of any pathogens they might carry. The importance of this procedure is demonstrated by the fact that 51% of the mice imported into TJL during the past three years carried pathogens. A large percentage of these infections are attributable to mouse hepatitis virus. All new strains are rederived by hysterectomy or embryo transfer to ensure a high health status and to avoid endangering existing strains. The process is described in the IMR section below.

## III. INFORMATION ON LABORATORY MOUSE STRAINS

TJL maintains a registry that lists all strains at the laboratory. A variety of reports are available to the scientific community, such as the Lane List of Named Mutations and Alleles of Polymorphic Loci of the Mouse, which lists spontaneous mutant genes and their genetic backgrounds. TJL publishes a *Handbook on Genetically Standardized JAX Mice* and a quarterly newsletter, *JAX Notes*, which contain information on mouse genetics and different types of strains. TJL's price list and product guide also include information on the genetics and uses of various types of mouse strains. Much of this information is accessible to researchers on-line through TJL's web page (http://www.jax.org). In the late 1970s a computer database cataloging genetic information on the mouse was started to complement the genetic resources at TJL. GBASE evolved into a comprehensive Mouse Genome Database (MGD) that has become a community database for mouse genomic information. MGD provides a comprehensive source of information on genetic mapping data, chromosome maps, specific gene typing for given strains, comparative mapping, and information on mouse DNA clones and probes. In addition, a Strain Characteristics Catalog with detailed information on strain characteristics is maintained as electronic files.[10]

Technical services specialists provide technical support and in-depth information on mice and their genetics to scientists and users of JAX® mice. This includes information on mouse models for studying human disease, genetic mapping, and other scientifically related issues. Special newsletters on topics, such as mouse models for neural tube defects, cataracts, skin diseases, and neurological mutations, are continually being prepared and updated. Written materials are available both through Animal Resources and the World Wide Web. This information is also distributed at TJL's exhibit booth displayed at scientific meetings throughout the year. Technical support is provided through the Customer Services Department's toll-free line (800-422-MICE) and through email (micetech@jax.org). Customer Service Representatives provide information on husbandry and availability of individual strains.

## IV. MOUSE MUTANT RESOURCE

The Jackson Laboratory holds the world's largest collection of spontaneous mouse mutants in its mouse mutant resource colonies. The majority of these mutants have been identified from the large breeding colonies at TJL where animal technicians are trained to recognize deviant phenotypic appearance and behavior. The identification and characterization of these mutants continues today as it has for the past 65 years. Until about 1980, the study of mammalian mutations relied almost entirely on the identification and characterization of mice with spontaneous mutations.

TJL's Mouse Mutant Resource (MMR) is a research center for analyzing new spontaneous mouse mutations and a resource for spontaneous mouse mutations and mutant stocks.[4] The three primary functions of the MMR are to (1) identify and characterize new mouse mutations for biomedical research, (2) publish information on these mutations to make them available to other scientists for biomedical research, and (3) propagate and distribute the mutant stocks with these mutations to scientists around the world. The research activities of the MMR are not restricted to a specific area and many different kinds of mutations are maintained and studied.

New spontaneous mutations are discovered through TJL's Phenotypic Deviant Search Program. Over 3.4 million mice were reared at the Animal Resources Production Facility in 1997. Each strain of either inbred or inbred-derived mice has unique identifiable characteristics that animal care technicians are trained to observe. Any marked departures from the visible norms observed in any mouse are recorded and the deviant mouse, its parents, and littermates are set out for biweekly clinics. Approximately 200 to 300 mice are selected each year as possible genetic deviants, and 100 to 200 of these are examined further by the scientific staff of the MMR and of TJL. The Production Department maintains a computerized list of all deviants, observed. This is likely the largest and most structured program of its kind in the world. It requires the coexistence of large breeding colonies of inbred mice, a highly trained animal technical staff, scientists with genetics and biological expertise, and research support services.

MMR personnel carry out genetic analysis and phenotypic characterization of new deviants that occur in TJL colonies. Genetic analysis involves determining mode of inheritance and mapping mutant genes to chromosomes. The method used to genetically map new mutations includes linkage crosses with DNA markers detected by Southern blotting and PCR analysis and with visible markers, such as coat color. A veterinary pathologist performs preliminary histopathological screening of all new mutants. The MMR has screening programs for hearing deficits, visual defects, and epilepsy-type brain wave patterns. MMR personnel frequently collaborate with scientists outside the MMR to complete analysis of specific kinds of mutants. Scientists interested in specific kinds of mutations are encouraged to visit the MMR and screen mutant mice for clinical symptoms of the particular disease in which they are interested. In addition to acquiring new mutant strains through the Phenotypic Deviant Search Program, the MMR imports spontaneous mutant stocks from outside investigators who are retiring or who cannot meet the demand for distribution from their research colonies.

Recessive mutations that produce lethality or sterility are maintained by special mating systems, usually progeny testing or ovary transplantation. Progeny testing means normal-appearing mice of unknown genotype (produced from heterozygotes mated together) are mated and their offspring observed. If affected mutants are produced, the breeding pair is considered "tested." If no affected progeny are produced in a total of 20, the breeding pair is discarded, or males are switched between two such matings. To maintain a strain by ovary transplantation, ovaries from homozygous mutant (*m/m*) females are transplanted into histocompatible or immune deficient (such as severe combined immune deficient *scid/scid*) hosts and bred to wild-type (+/+) males. The resulting progeny are obligate heterozygotes and are intercrossed to produce more homozygous females. A coat-color coding system is used when possible. For example, many mutations in the MMR are maintained on the C57BL/6J (B6) (*a/a*, black) genetic background or on a C3H (C3) substrain fixed for nonagouti (*a/a*, black). Ovaries are transplanted into B6C3 F1 hybrid hosts whose genotype at the agouti locus is agouti (*A/A*) or white bellied agouti (*A/A*ʷ). When the *a/a* ovaries are mated to a male who is *a/a*, all pups from the donor (mutant) ovary will be black. If host ovary tissue is not entirely removed, pups from the host ovary will be agouti. Increasingly, as spontaneous mutations are cloned, these strains are being maintained by genotyping breeders using DNA typing methods.

Although recombinant DNA technology has made it possible to engineer specific types of mutations into targeted genes, spontaneous mutations identify novel genes and provide a source of mutations in genes that have not yet been cloned. Advances in molecular technology make it possible to identify the mutated gene by positional cloning or identifying a mutation in a closely linked candidate structural gene.

## V. INDUCED MUTANT RESOURCE

Mutant mice also have been created by random mutagenesis using radiation,[11] chemical mutagens such as N-ethyl-N-nitrosourea (ENU),[12] gene trapping with retroviral vectors,[13,14] and genetic engineering.[15-19] As early as the 1940s, risk assessment studies were contributing radiation and chemically induced mutations in mice. Since 1980, genetic engineering has become a powerful tool and the most common and rapidly growing type of induced mutant mice.

In 1980, the first genetically engineered mice were produced when it was shown that DNA microinjected into the male pronucleus of a fertilized egg became randomly incorporated into the host genome, usually in multiple copies.[15,16] These transgenic mice had new genetic material added to their genomes and, since this could include human genes, this discovery initiated a new wave of research into mammalian gene function. A second major advancement occurred in the mid and late 1980s when techniques were described for the deliberate creation of mutations in known genes, called targeted mutations or "knockouts."[17-19] Advances in recombinant DNA technology have made it possible to engineer specific alterations into cloned genes. These altered genes are then introduced into embryonic stem cells (ES cells), and homologous recombination inserts specifically mutated genes into chromosomes in place of the endogenous gene in a small percentage of the ES cells. Positive and negative selection procedures aid in cloning correctly targeted ES cells.[17-19] Finally, the mutation-bearing ES cells are microinjected into blastocysts to produce chimeric animals. If the altered gene finds its way into the germline, it will be transmitted to the offspring of the founder chimera. Transgenic technology makes it possible to add new genetic material to the mouse genome; targeted mutations make it possible to remove genetic functions by inactivating individual genes. Conditional targeted mutations allow temporal and tissue-specific inactivation of genes.[20-22]

Genetically engineered mice provide mouse models for specific human disorders and open entirely new lines of research into the physiological functions of many genes, including those contributing to such common diseases as cancer, cardiovascular disease, Huntington's disease, cystic fibrosis, diabetes, and many congenital defects.[23] Even for diseases that we do not conventionally think of as genetic, such as AIDS, the new mouse models are important for exploring the pathophysiology of such diseases and may be used as test systems for developing new therapeutic interventions, including gene therapy and other more conventional strategies.

The six functions of the IMR are to (1) select and (2) import mutant strains into the resource, (3) cryopreserve embryos and gametes as insurance against loss and for efficient maintenance, (4) develop mutant strains with defined genetic backgrounds, (5) maintain and distribute strains to other scientists and (6) provide information on the strains maintained in the IMR. Each of these functions is discussed briefly below.

## A. SELECTION OF MUTANTS

There has been an explosion in the rate of production of transgenic and targeted mutation mice since the two technologies were developed (Figure 13.1). It is not possible (or necessary) for TJL to import and distribute every genetically engineered mouse produced. Therefore, TJL's Genetic Resource Committee has established criteria for selecting mutant strains: (1) the immediate need for use in biomedical research, (2) the number of requests for mice being received by the investigator(s) who created them, (3) the potential for future research, (4) the time and effort needed to replace or recreate the mutant, and (5) the uniqueness of the mutation. An external Advisory Board and Associated Boards have been established to assist in the selection process and reflect community needs. Potential importations into the IMR are identified by (1) requests initiated by the creator of the mutant, (2) proposals from the IMR supervisor based on literature scans, attendance at scientific meetings, contact with healthcare agencies, and suggestions from investigators inquiring about specific mutants, (3) suggestions by the scientific staff at TJL, and (4) suggestions from the Advisory or Associated Boards.

## B. IMPORTATION OF MUTANT STOCKS

Mice arriving at TJL are isolated and quarantined in isolators in a dedicated facility. Sample mice are subjected to microbial and viral evaluations; others are mated and their progeny rederived by hysterectomy. The derived mice are raised by foster mothers with defined flora, and, after pups are weaned, the foster mothers are tested for specific pathogens before the weaned litters are transferred to strict barrier mouse rooms. The importation of mice can be a costly and time-consuming procedure. The time required for importation depends on the genotype and number of mice supplied for the rederivation, individual breeding and strain characteristics, and the generation time of the mouse. A strain that breeds well usually requires 10 to 12 weeks before

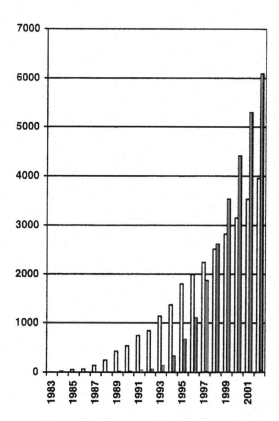

**FIGURE 13.1** The number of publications citing either transgenic (open bars) or targeted mutation mice (shaded bars). Current (1985–1997) values were obtained using SilverPlatter to search MEDLINE. Values for 1998–2002 were projected utilizing the historical rates and trends for transgenic publications. The majority of publications citing targeted mice represent newly created strains.

the first litter can be released from importation. A strain that does not breed well may take six months or longer before a litter can be released. After clearing importation, a strain is expanded to a colony size sufficient to support limited distribution. Again, the generation time is the rate-limiting event. Even though mice have a short gestation time for a mammal, their generation time is much longer than that of *C. elegans or Drosophila*. It may require six months to one year from the time of arrival of a new strain at TJL before animals can be distributed as breeding pairs. This interval is a necessary requirement to ensure the health of the animals being distributed and the subsequent health of the colonies into which they are received.

## C. CRYOPRESERVATION OF EMBRYOS AND GAMETES

Embryos from every strain in the IMR are cryopreserved once the strain is established. The process of cryopreservation is described later in the Cryopreservation

Resource section. The number of mutation-bearing alleles frozen for each strain depends on whether it is undergoing strain development (backcrossing the mutation onto an inbred strain). For mutants arriving at the laboratory on a segregating genetic background and undergoing strain development, 150 mutation-bearing embryos are frozen initially from the segregating background. An additional 500 embryos from this strain are frozen upon completion of the backcrossing. The initial freezing of 150 embryos from the segregating strain is a precaution to prevent accidental loss of the mutation during the backcrossing, which may take as long as two years. For mutants arriving at TJL on an inbred background, 500 mutation-bearing embryos are frozen. More recently we have also begun to utilize sperm cryopreservation when the mutation is on a genetic background that allows this.

## D. Strain Development

The majority of the targeted mutation mice, as well as many of the transgenic animals, arrive in the IMR on segregating genetic backgrounds, a result of the technology involved in their production. The usefulness of these mice is greatly improved if the mutation is transferred to an inbred genetic background. For this reason, new mutations or transgenes are often backcrossed onto an inbred strain, usually C57BL/6J. Additional genetic backgrounds are considered depending on their potential usefulness to researchers. This strain development occurs concomitantly with the distribution of the mutant on the segregating background. Many useful research models may be generated by backcrossing certain targeted mutations or transgenes onto inbred strains with specific characteristics. For example, crossing a tumor suppressor gene knockout, such as *Trp53* (p53), onto an inbred strain with specific tumor susceptibility may be useful in determining the role of *Trp53* in tumorigenesis for that particular strain or tumor. Likewise, certain double knockout or transgenic strains may be generated to study the specific roles of these genes. The IMR is developing a research program to generate and characterize these mutant strains created by breeding. As soon as the characterization of these new strains is completed, they are made available to the scientific community. To make mice from these congenic strains rapidly available mice are offered after five generations (N5) of backcrossing, while backcrossing is continued to at least N10. N5 mice are statistically >95% host genetic background and can be used for most purposes.

## E. Maintenance and Distribution

The rapid distribution of transgenic and targeted mutant strains to the scientific community is a primary objective of the IMR. Distribution of animals from the IMR begins as soon as the mutants have cleared the importation process and the colony has been expanded to a size large enough to support limited distribution of mating pairs without jeopardizing the breeding colony. This occurs regardless of the genetic background of the mutant. During this initial period, single requests for large numbers of mice are supplied in smaller groups or "lots" interspersed with shipments to other investigators so that the resources are distributed equitably. The colony is expanded as rapidly as reproductive characteristics of the strain will allow to a size large enough to support anticipated demand. Colony expansion concurrent with limited distribution is a time-consuming process. Approximately 10 weeks are required from the time mating pairs are set up until the distribution of six-week-old animals can begin. Every effort is made to make these mice available as soon as possible, but a waiting period may be necessary for those strains in high demand. For mutant or transgenic mouse strains undergoing strain development, distribution on the mixed background continues until the backcrossing is completed, at which time mutants on the inbred strain are made available. The genetic backgrounds and genotypes of all mice distributed are clearly indicated on the shipping tags.

Approximately 30% of the genetically engineered mutants that have been accepted into the IMR are homozygous lethals or do not survive long enough to produce progeny. Genotypes of all offspring in these strains must be identified prior to distribution. Mice from these strains are supplied as genotyped (tested) heterozygotes. PCR (polymerase chain reaction) is the method of choice to genetically type DNA from these animals.

Distribution of animals from the IMR follows the same guidelines as distribution of mice from any TJL colonies. In general, mice are distributed on a first come, first served basis. TJL places no restrictions on the subsequent breeding of the mice distributed, with the exception that they or their offspring may not be bred for resale, are to be used solely for research purposes, and may not be transferred outside the recipient's institution. There are sublicensing requirements imposed by the originating institution for some of the IMR animals, directed at commercial or for-profit companies. Strains requiring sublicensing are identified in all IMR literature and in the price list. If a sublicense is required, it must be obtained by the investigator from the institution where the mutants originated. In some cases, legal (licensing) negotiations may delay the distribution of a strain.

## F. INFORMATICS

The IMR maintains an on-line database that provides information on strains available from the IMR, including brief descriptions, how the mutation or transgene was created, genetic background, animal husbandry, genotypes available, genotyping protocols, initial references, and price: URL: http://www.jax.org.resources. The IMR database also links to The Mouse Genome Database (MGD)—URL: http://www.informatics.jax.org—for descriptive information on genes. In addition, TJL recently assumed responsibility for TBASE, a database for transgenic and targeted mutant animals originally developed by Dr. Richard Woychik.[24] TBASE contains information on, among other items, the correct nomenclature, biology, and literature for transgenic and targeted mutants, and covers stocks not available at TJL.

## VI. INBRED, CONGENIC, AND RECOMBINANT INBRED STRAINS

*Inbred strains* are produced by more than 20 generations of sibling matings. Although strains are defined as inbred after 20 generations, even after 40 generations, heterozygosity may still exist but is expected to be <1%.[25] Many inbred strains at TJL have been inbred for more than 100 generations. Inbred strains from TJL are available primarily from the Production and Distribution Colony. To control subline divergence within individual inbred strains, the Foundation Stocks Unit was established in 1959. This program contains the pedigreed source breeding pairs for strains of mice produced in the Animal Resources Production and Distribution Unit. Strains are expanded in the Pedigreed Expansion Stocks Colony from breeders provided by Foundation Stocks. High demand strains are further expanded in the Production and Distribution Colony. With rare exceptions for strains with low breeding performance, strains are not expanded more than 10 generations of breeding beyond the Foundation Stocks pedigreed line.

*Congenic strains* are created by successive backcrosses in which one strain (the donor) donates a segment of chromosome containing a gene of interest to the recipient (background or host) strain. Usually 10 to 12 backcross generations are carried out. Congenic strains are genetically almost identical to the background strain except for a short chromosomal segment contributed by the donor strain.[26]

*Recombinant inbred (RI) strains* are sets of inbred strains derived by crossing mice from two defined parental strains, intercrossing the F1 progeny and then establishing lines by making sibling matings beginning with mice of the F2 generation.[8] This mating system results in sets of related inbred substrains that have different combinations of the original parental genomes. RI strain sets are useful for genetically mapping phenotypic or quantitative traits that differ between the parental strains. Since all mice in each individual RI strain of a set are considered genetically identical, several mice can be typed to classify a quantitative trait. Congenic and recombinant inbred strains are maintained in the Special Mouse Stocks Resource (SMSR) within the Production and Distribution Unit.

## VII. CYTOGENETIC MODELS RESOURCE

The Cytogenetic Models Resource contains stocks carrying Robertsonian chromosomes, reciprocal translocations, and a segmental trisomy. The breeding colony in the Cytogenetic Models Resource focuses on chromosome aberrations that can be used to study aneuploidy for mouse Chromosome 16, in which many human Chromosome 21 genes are conserved. Included are selected reciprocal translocations, Robertsonian chromosome stocks that can be used to produce embryos with trisomy for the entire Chromosome 16, and a segmental trisomy, Ts(17[16])65Dn, for the segment of mouse Chromosome 16 that is homologous to human Chromosome 21, trisomic in Down syndrome. The Resource also has several stocks of mice transgenic for genes from human Chromosome 21 or their mouse homologs. A large number of Robertsonian chromosome stocks previously maintained in this Resource is now available from the Cryopreservation Resource. Each mouse chromosome is present in at least two different Robertsonian chromosomes in these strains so that embryos trisomic for each of the 19 mouse autosomes can be produced.

## VIII. CRYOPRESERVATION RESOURCE

Cryopreservation of germplasm is widely recognized as the best means to ensure against loss of valuable strains and efficiently maintain strains when they are not in immediate demand. In the frozen state, embryos are unaffected by diseases, sudden reproductive failure, certain environmental accidents and genetic "contamination" that can threaten breeding colonies. Reconstitution of valuable strains that have been lost because of reproductive failure or genetic contamination has occurred nine times in the past 15 years at TJL. In addition, strains that are no longer in demand may be maintained as frozen embryos, thus conserving expensive mouse room space. Experience has shown that strains can be maintained in less space or can be removed from conventional breeding altogether when frozen embryos are available as a backup. There is no evidence to indicate that genetic mutations accumulate during storage of frozen embryos and, likewise, there is no evidence to indicate that the length of time of storage in liquid nitrogen impairs the ability to recover embryos.[27] Strains whose embryos were cryopreserved for 15 years have been recovered with no unusual difficulty. Mouse embryos were first successfully frozen in 1972.[28,29] More recently it has become possible to cryopreserve mouse sperm and ovaries, a more efficient procedure for freezing mutations than embryo freezing provided the genetic background does not have to be recovered. Because sperm cryopreservation is not yet reliable for most inbred strains, however, this method cannot yet be used to preserve nonstandard genetic backgrounds. It is anticipated that current research into sperm cryopreservation will result in this method eventually being the primary method for the protection and preservation of mouse strains. If successful, sperm cyropreservation will reduce the time and expense currently required for embryo cryopreservation. The Cryopreservation Resource provides assurance against accidental loss of strains maintained in research and genetic resource colonies, and serves as an archive for strains not maintained in breeding colonies.

The standard procedure at TJL is to cryopreserve a minimum of 500 mice from nonmutant strains and 500 mutation-bearing embryos from mutant strains. Depending on the mating scheme used, the embryo freezing may require 500 to 1000 or more embryos, the number depending upon expected embryo genotype. The banked embryos are stored at two physical locations as protection against a catastrophic loss of either facility. Between 100 and 200 embryos from each strain are stored in the facilities of the National Institute of General Medical Sciences Human Genetic Mutant Cell Repository, Camden, New Jersey. The remainder are stored at TJL. Strains are reconstituted from frozen embryos or sperm only when mice from those strains cannot be obtained from other sources. Over 100 strains are reconstituted each year to fill orders from approximately 800 strains that are maintained only in the frozen state.

Over 1600 different strains of genetically defined laboratory mice are now preserved as frozen eight-cell embryos, sperm, or ovaries stored in liquid nitrogen storage containers at TJL. In addition to backup, other benefits are the reduction in the number of different strains or size of colonies maintained by relatively more costly conventional breeding procedures and the retardation of the genetic change that occurs through time in an inbred strain as a result of accumulated mutations (genetic drift).

## IX. DNA RESOURCE

The Mouse DNA Resource preserves and distributes genomic DNA from TJL's strains to other scientists. The DNA is high-molecular-weight DNA obtained by phenol–chloroform extraction from the tissues of individual mice and is suitable for Southern blotting and amplification by the polymerase chain reaction. DNA is usually extracted from male mice to recover both X and Y chromosome DNA. DNA from remutations at known loci is preserved as well to provide a resource for gene cloning experiments. If a candidate gene for a mutation is found, DNA from the extinct mice carrying the remutation is available for analysis.

## ACKNOWLEDGMENTS

The MMR has been continuously supported by the National Science Foundation (NSF) since 1960 (currently DBI 95-0222) and by the National Center for Research Resources (NCRR) at the National Institutes of Health (NIH) since 1970 (P40 RR01183). It also receives funding from the National Institute of Child Health and Human Development (NICHD; HD53230), the National Eye Institute (NEI; R01 E05578), the National Institute for Deafness and Communicative Disorders (NIDCD; DC62108), and the Foundation Fighting Blindness. The IMR is currently supported by the NCRR at the NIH (P40 RR RR09781, P40 RR11081) and the Howard Hughes Medical Institute (grants 76193-502402, 76196-502403). Previous support has come from The March of Dimes Birth Defects Foundation (grant TY92-1314), the American Cancer Society (grant RD-366), the American Heart Association, the National Multiple Sclerosis Society, the Amyotrophic Lateral Sclerosis Foundation, and the Cystic Fibrosis Foundation (grant 5901). This support provides for the importation, cryopreservation, and strain development of accepted strains, as well as database development and maintenance. Strain maintenance and distribution costs are supported from the sale of mice. The cytogenetic resource is supported by the National Institute for Child Health and Human Development (HD73265). The cryopreservation resource is supported by the National Center for Research Resources (P40 RR01262). All four resources also are supported by revenues generated by the distribution of mice.

## REFERENCES

1. Andersson, L., Archibald, A., Ashburner, M., Audun, S., Barendse, W., Bitgood, J., Bottema, C., Broad, T., Brown, S., Burt, D., Charlier, C., Copeland, N., Davis, S., Davisson, M., et al., Comparative genome organization of vertebrates: The First International Workshop on Comparative Genome Organization, *Mamm. Genome*, 7, 717, 1996.
2. Davisson, M. T., Lalley, P. A., Peters, J., Doolittle, D. P., Hillyard, A. L., and Searle, A. G., Report of the comparative subcommittee for human, mouse, and other rodents (HGM11), *Cytogenet. Cell Genet.*, 58, 1152, 1991.
3. Mouse Genome Database (MGD), Mouse Genome Informatics, The Jackson Laboratory, Bar Harbor, Maine (URL:http//www.informatics, jax.org/), 1998.

4. Davisson, M. T., The Jackson Laboratory Mouse Mutant Resource, *Lab Anim.*, 19, 23, 1990.
5. Sharp, J. J. and Davisson, M. T., The Jackson Laboratory Induced Mutant Resource, *Lab Anim.*, 23, 32, 1994.
6. Sharp, J. J. and Mobraaten, L. E., To save or not to save: The role of repositories in a period of rapidly expanding development of genetically engineered strains of mice, in *Transgenic Animals—Generation and Use*, Houdebine, M., Ed., Harwood Academic Publishers GMBH, Switzerland, 1997, 525.
7. Snell, G. D., Methods for the study of histocompatibility genes, *J. Genet.*, 49, 87, 1948.
8. Bailey, D., Recombinant inbred strains and bilinial congenic strains, in *The Mouse in Biomedical Research*, Foster, H. L., Small, J. D., and Fox, J. G., Eds., Academic Press Inc., New York, 1981, 223.
9. Mobraaten, L. E., The Jackson Laboratory Genetics Stocks Resource Repository, in *Frozen Storage of Laboratory Animals*, Zeilmaker, G. H., Ed., Gustav Fischer, Stuttgart, 1981, 165.
10. Eppig, J. T., Blake, C. S., Bradt, D. W., Grant, P., Guidi, J. N., Hillyard, A. L., Jones, L. M., Kosowsky, M. R., Maltais, L. J., Ormsby, J. E., Rockwood, S. F., Snell, T. C., and Volmer, M. S., The mouse genome database, *The Sixth International Mouse Genome Conference*, Buffalo, NY, Abst. 28, 1992.
11. Searle, A. G., Mutation induction in mice, *Adv. Radi. Biol.*, 4, 131, 1974.
12. Moser, A. R., Pitot, H. C., and Dove, W. F., A dominant mutation that predisposes to multiple intestinal neoplasia in the mouse, *Science*, 247, 322, 1990.
13. Friedrich, G. and Soriano, P., Insertional mutagenesis by retroviruses and promoter traps in embryonic stem cells, *Meth. Enzymol.*, 225, 681, 1993.
14. Zambrowicz, B. P., Friedrich, G. A., Buxton, G. C., Lilleberg, S. L., Person, C., and Sands, A. T., Disruption and sequence identification of 2,000 genes in mouse embryonic stem cells, *Nature*, 392, 608, 1998.
15. Gordon, J. W., Scangos, G. A., Plotkin, D. J., Barbosa, J. A., and Ruddle F. H., Genetic transformation of mouse embryos by injection of purified DNA, *Proc. Natl. Acad, Sci. U.S.A.*, 77, 7380, 1980.
16. Wagner, T. E., Hoppe, P. C., Jollick, J. D., Scholl, D. R., Hodinka, R. I., and Gault, J. B., Microinjection of a rabbit ß-globin gene into zygotes and its subsequent expression in adult mice and their offspring, *Proc. Natl. Acad. Sci. U.S.A.*, 78, 6376, 1981.
17. Smithies, O., Gregg, R. G., Boggs, S. S., Koralewski, M. A., and Kucherlapati, R. S., Insertion of DNA sequences into the human chromosomal ß-globin locus by homologous recombination, *Nature*, 317, 230, 1985.
18. Thomas K. R., and Capecchi, M. R., Site-directed mutagenesis by gene targeting in mouse embryo-derived stem cells, *Cell*, 51, 503, 1987.
19. Mansour, S. L., Thomas, K. R., and Capecchi, M. R., Disruption of the proto-oncogene *int-2* in mouse embryo-derived stem cells: a general strategy for targeting mutations to non-selectable genes, *Nature*, 336, 348, 1988.
20. Gu, H., Marth, J. D., Orban, P. C., Mossmann, H., and Rajewsky, K., Deletion of DNA polymerase beta gene segment in T cells using cell type-specific gene targeting, *Science*, 265, 103, 1994.
21. Kuhn, R., Schwenk, F., Aguet, M., and Rajewsky, K., Inducible gene targeting in mice, *Science*, 269, 1427, 1995.

22. Kistner, A., Gossen, M., Zimmermann, F., Jerecic, J., Ullmer, C., Lubbert, H., and Bujard, H., Doxycycline-mediated quantitative and tissue-specific control of gene expression in transgenic mice, *Proc. Natl. Acad. Sci. U.S.A.*, 93, 10933, 1996.
23. Paigen, K., A miracle enough: the power of mice (Review), *Nature Med.*, 1, 215, 1995.
24. Woychik, R. P., Wassom, J. S., Jacobson, D. A., and Kingsbury, D. T., TBASE: a computerized database for transgenic animals and targeted mutants, *Nature*, 363, 375, 1993.
25. Green, E. L., *Genetics and Probability in Animal Breeding Experiments*, Macmillan Publishers Ltd., London, 1981, 141.
26. Snell, G. D. and Bunker, H. P., Histocompatibility genes of mice., V, Five new histocompatibility loci identified by congenic resistant lines on a C57BL/10 background, *Transplantation*, 3, 235, 1965.
27. Lyon, M. F., Glenister, P. H., and Whittingham, D. G., Long-term storage of frozen mouse embryos under increased background irradiation, in *Freezing of Mammalian Embryos*, Ciba Foundation Symposium 52 (new series). Amsterdam, Elsevier, 1977, 273.
28. Wilmut, I., The effect of cooling rate, warming rate of cryoprotective agent, and stage of development on survival of mouse embryos during freezing and thawing, *Life Sci.*, 11, 1071, 1972.
29. Whittingham, D., Leibo, S., and Mazur, P., Survival of mouse embryos frozen to -196° and -269°C, *Science,* 178, 411, 1972.

# Index

## A

Abdominal tissues, 72
Accession number, 50–51
Acetone, 124
Acetonitrile, 124
Acrolein, 123
Adrenal glands, 76, 81, 82
AEC, *see also* 3-Amino-9-ethylcarbazole
Agouti locus, 6
Alkaline phosphatase, 135, 174
Allelism, 18
AMCA, *see* 7-Amino-4-methyl-coumarin-3-acetic acid
3-Amino-9-ethylcarbazole (AEC), 136, *see also* Chromagens
7-Amino-4-methyl-coumarin-3-acetic acid (AMCA), 139
Analog technology, 92, *see also* Photography
Ancestry information, 44
Animal models, comparative pathology, 102–103
Antibody
    fluorescent-labeled, 138
    nonspecific binding, 139
    sources for immunochemistry/immunofluorescence methods, 132–133, 134
Anus, 75
Asphyxiation, 68–69
Autofluorescence, 139
Autofocus cameras, 92
Autoradiography, 156–162
Avidin–biotinylated enzyme complex, 133, 136

## B

B5 fixative, 65, 66
BAC, *see* Bacterial artificial chromosome
Backcross (BC) generation
    colony establishment, 5
    cryopreservation of embryos and gametes at Jackson Laboratory, 184
    genetic mapping, 18, 20, 21, 22, 23
    linkage analysis and mutation assignment to chromosome interval, 27, 28

Background staining, 139, 174
Backgrounds, photography, 94, 95–96
Bacterial artificial chromosome (BAC), 31
Bar codes, 51
Barbiturates, 69
Barriers, physical, 96
BC, *see* Backcross generation
BCIG, *see* 5-Bromo-4-chloro-indolyl-ß-d-galactopyranoside
BCIP, *see* 5-Bromo-4-chloro-3-indolyl phosphate
BCIP/IBT, *see* 5-Bromo-4-chloro-3-indolyl phosphate/iodoblue tetrazolium
BCIP/NBT, *see* 5-Bromo-4-chloro-3-indolyl phosphate/nitroblue tetrazolium
Behavior, patterns, 59
Biological markers, 6
Biotin, 133
Biotin-UTP probes, 167
Biotinylated antibody, 133, *see also* Antibody
Black cloth backgrounds, 94
Blood, 61–62, 70
Bouin's solution, 65, 66
Boxes, three-sided, 97
Brain, 77–78, 84, 85
BrdU, *see* Bromodeoxyuridine
Breeding
    colonies losses and cryopreservation, 187
    colony establishment, 3
    schemes, 19–22, 29
5-Bromo-4-chloro-indolyl-ß-d-galactopyranoside (BCIG), 136, *see also* Chromagens
5-Bromo-4-chloro-3-indolyl phosphate (BCIP), 171, 174
5-Bromo-4-chloro-3-indolyl phosphate/iodoblue tetrazolium (BCIP/IBT), 136, *see also* Chromagens
5-Bromo-4-chloro-3-indolyl phosphate/nitroblue tetrazolium (BCIP/NBT), 136, *see also* Chromagens
Bromodeoxyuridine (BrdU), 59, 112, 113
Brother–sister matings, 4, *see also* Matings
Buffer systems, 124, 126, 147, 171, *see also* Individual entries